David Demaree

GIT FOR HUMANS

MORE FROM A BOOK APART

Going Responsive
Karen McGrane

Responsive Design: Patterns & Principles
Ethan Marcotte

Designing for Touch
Josh Clark

Responsible Responsive Design
Scott Jehl

You're My Favorite Client
Mike Monteiro

On Web Typography
Jason Santa Maria

Sass for Web Designers
Dan Cederholm

Just Enough Research
Erika Hall

Content Strategy for Mobile
Karen McGrane

Design Is a Job
Mike Monteiro

Mobile First
Luke Wroblewski

Visit abookapart.com for our full list of titles.

Publisher: Jeffrey Zeldman
Designer: Jason Santa Maria
Managing Director: Katel LeDû
Editor: Caren Litherland
Technical Editor: David Yee
Copyeditor: Katel LeDû
Proofreader: Caren Litherland
Compositor: Rob Weychert
Ebook Producer: Ron Bilodeau

ISBN: 978-1-9375573-9-3

A Book Apart
New York, New York
http://abookapart.com

10 9 8 7 6 5 4 3 2 1

TABLE OF CONTENTS

FOREWORD

LET ME TELL IT to you straight: Git is infuriating.

Wait! Don't run off just yet. Because while Git *is* infuriating, it's also critical in two very different and equally compelling ways: first, speaking practically, Git is a prerequisite for collaborating on websites or applications, which, if you're holding this book, is probably something you are wont to do. And second, Git is a kind of model for present-day collaboration—that is, collaboration among distributed teams, working asynchronously, on a shared body of work.

So while you don't have to love Git, you do have to *know* it.

Many Git tutorials bend over backwards to map Git's arcane practices on to real-world phenomena, often leaving readers hanging from trees wondering which branch is about to snap. Here, David Demaree dispenses with that nonsense, inviting you to learn about Git on its own terms. Rather than extended, creaking metaphors, Demaree patiently explains in plain language the core principles underlying Git that every designer, developer, content strategist, and product manager needs to know. The result is a brisk, clear book you can read in a few hours and then return to your terminal, ready to confidently pull and merge.

But there's more here, and you'd be wise not to miss it. Along with the commands and syntax, there's keen advice in these pages about working on a team. Knowing when and how to commit a change is more than just a means of updating code—it's also a practice for communicating and sharing work. It's a process, and a remarkably powerful one. So while Git's quirks often leave newcomers reaching for drink, its influence on people who make websites is well deserved. By all means, devour the following chapters in order to understand how to manage merge conflicts and interpret a log. But don't forget that Git's ultimate audience isn't machines—it's humans.

—Mandy Brown

INTRODUCTION

WHEN I STARTED making websites as a hobby in 1995, being a web developer meant knowing HTML. That's it. Neither JavaScript nor CSS would ship in browsers for a year, and Flash wouldn't exist until later in the decade. The web was just starting to become a rich medium full of engrossing content, and anyone with a text editor who could remember a dozen or so tags could participate. It was nice.

Twenty years later, web development is no longer so simple. HTML, CSS, and JavaScript remain the foundation of our work, but over their history—their recent history in particular—they've evolved from languages for crafting documents, simple enough that most designers could write them from memory, into a platform for writing applications. It feels like we don't make web *pages* anymore; we make *themes* or *templates* or, if we're really ambitious, we make *apps*. We're producing thousands of lines of increasingly complex code, and we're sharing responsibility for managing that code with more people, in more and more ways. We have the power to make truly amazing things for our users, things we never could have imagined when the web was young—but at the cost of feeling like gerbils running on a technology treadmill.

Frank Chimero put it well (http://bkaprt.com/gfh/00-01/):

Now is the time to come clean: GitHub is confusing, Git is confusinger, pretty much everything in a modern web stack no longer makes sense to me, and no one explains it well, because they assume I know some fundamental piece of information that everyone takes for granted and no one documented, almost as if it were a secret that spread around to most everyone sometime in 2012, yet I somehow missed, because—you know—life was happening to me, so I've given up on trying to understand, even the parts where I try to comprehend what everyone else is working on that warrants that kind of complexity, and now I fear that this makes me irrelevant, so I nestle close to my story that my value is my "ideas" and capability to "make sense of things," even though I can't make sense of any of the above—but really, maybe I'm doing okay, since it's all too much to know. Let the kids have it.

Git is hardly the most complicated new web technology, but it's a part of this new stack that all parts of the stack have in common. You cannot escape Git if you want to participate in the new platform-y web. At some point you'll need to contend with it, either directly or as a transport mechanism used by some other tool. And that may very well be why Git is a poster child for this sea change in how we make websites.

Plenty of books, blog posts, and other online materials have cropped up to teach users at all levels how to use Git. Yet despite this wealth of tutorials, some days it feels like you can't turn around without bumping into someone complaining that Git makes no goddamned sense. And yet we use it. It seems like we have to use it, despite fearing that we cannot confidently use it, leaving us to feel like we're running around the house with a big pair of scissors.

And it's not just designers like Frank Chimero. Folks who are new to the web, or who want to work in fields only tangentially related to web development (like writing or open data), are also forced to live on a Git planet, as are tons of us who like the engineered web just fine, but who still feel flummoxed by Git.

Having spent most of the last decade using Git on almost every project, delving at times into some of the darkest, weirdest corners of Git behavior, I can safely say that *it's not you, it's Git*. Git isn't difficult because you're not smart enough, or because you missed an important meeting. Git is difficult because *Git is difficult*.

Git is difficult, in part, because it embodies what Joel Spolsky calls a "leaky abstraction" (http://bkaprt.com/gfh/00-02/). Abstractions, in the software sense, are things that make a task conceptually easier to handle by covering up the elements that make them hard. Interfaces are abstractions: there's absolutely no relationship between dragging a file to a trash can icon to delete it and actually removing the file from your hard drive, except that a designer thought it would make the concept of deleting files easier to understand. And it works! Even though I used computers for more than a decade without knowing how the shift key worked (seriously!), I never had trouble understanding how to get rid of a file.

Abstractions are there to protect us from complexity. A *leaky* abstraction fails at its job by letting some of the underlying complexity peek through, the same way a leaky umbrella fails at its job of keeping you dry. To quote one of Spolsky's examples:

> *You can't drive as fast when it's raining, even though your car has windshield wipers and headlights and a roof and a heater, all of which protect you from caring about the fact that it's raining (they abstract away the weather), but lo, you have to worry about hydroplaning... and sometimes the rain is so strong you can't see very far ahead so you go slower in the rain...*

Git's interface is "leaky" because its command-line interface fails to protect you, the user, from knowing how it works under the hood. And one reason why Git is so scary, is that it has its own internal logic, which doesn't always map to how we humans are used to organizing information. So knowing how it works is sometimes essential to getting it to work. To use Git successfully, you sometimes need to be able to apply Git logic to situations where human logic (and Git's supposedly human-friendly abstractions) fails. In other words, to master Git, you have to think like Git.

I want to help you understand how Git thinks.

Believe it or not, Git's challenging conceptual model is a feature, not a bug. Using Git feels like running with scissors because it's a powerful tool that will let you bend time and space to your will, which sounds like—and is—a lot of responsibility to put in the hands of mere humans. But Git believes in your ability to handle such might, and so do I.

Let's get started.

1
THINKING IN VERSIONS

IF YOU'VE BEEN AROUND FOR A WHILE, you may recall that it was once common for authors to carve their words into stone. Leaving aside the stamina this required, the size, weight, and expense of the material made it somewhat inconvenient for writers to make changes to their work once it was completed. Fixing a typo, let alone clarifying one's message or making the language flow better, required cutting away sections of stone or finding a different stretch of cave wall to write on. Even on the rare occasion when it was truly necessary to change something, it was hard—physically hard—to hold on to old revisions, making it almost impossible to compare the finished version of a poem, recipe, or cave painting with the version that preceded it, or to experiment with alternate drafts. Writing anything down at all seemed like magic.

Over the centuries, it became easier to put things in writing, which in turn made it much easier for writers to explore different approaches or to change their minds, both during the creative process and after the fact. This had the added perk of making ideas and language easier to disseminate, which made just about everyone (at least potentially) a writer.

But until the introduction of computers, the best way to record or distribute an idea was still to inscribe it on a physical object, like a piece of paper, which took time and cost money. The expense of making additional versions made it so that creating a first draft of anything—a novel, a blueprint, a painting, even a photograph—had an air of finality about it.

Many of us still think this way about how we produce our work. Taking the time to clarify and improve something over the course of multiple drafts feels like a luxury. Computers and networks have made it infinitely cheaper to spread information, but iteration still requires two things: time and discipline.

When I was in school, a few teachers attempted to show me and my fellow students the value of iteration (not to mention starting projects earlier than the night before they were due) by asking us to turn in not only our final papers, but also the drafts that preceded them. Rather than turn us into nimble iterative thinkers, though, mostly this kept us up late scrambling to meet deadlines a few times a semester instead of just once. Given the choice between spending extra time making three versions of an essay versus one—even though in doing so we'd make each one better than the last—we'd much rather be lazy, settle for flawed or mediocre work, and spend our time catching up on old episodes of *Fringe*.

But there are at least two areas of our written culture where making incremental changes, and tracking those changes across multiple versions, is not just helpful but crucial: law and (more important for our story) software source code.

Like other kinds of writing, source code went through what might be called an analog phase. Early computers had to be programmed by punching holes into cards, which were fed individually into the machines, which in turn performed the instructions encoded into the cards and returned a result. (The computing words "bugs" and "debugging" are popularly attributed to Grace Hopper, who traced problems in the operation of the Harvard Mark II computer to moths that had nested among the data relays.)

Early coders endured some of the same problems that early writers did: making changes on physical media like punch cards was time-consuming and expensive. Their programs took hours or days to run, and an error in a program meant that the whole sequence needed to be restarted from scratch, making it very important to get it right the first time whenever possible.

Computing languages have to be understood by machines, which—science fiction notwithstanding—remain much stupider than we are. Where a human can read "their" instead of "they're" (or the code equivalent) and just sigh at an author's poor attention to detail, a computer will crash. A computer system that crashes is not very useful, so software makers tried to make things easier the best way they knew how: by building more systems.

ELEMENTS OF VERSION CONTROL

What they came up with is a *version control system*. The basic principle of version control is this: instead of keeping only the latest copy of something, you hold on to each successive revision as you work, so that you can refer or revert back to an older version if you need to. Although you can use software tools—one of which is the subject of this book—to help you, version control is more importantly a *practice*. It's something we *do*—not just the tools we use to do it.

Many of us have had projects where we kept copies of old versions of our work, saving new versions by using an app's Save As... command to give each new copy its own name. Perhaps we marked the new filename with the current date (project_2014-04-15.doc), or maybe we added a version number (mockup-1a.psd). Both are rudimentary, but entirely valid, forms of version control.

Version control systems like Git work by keeping a copy of each successive version of your project in something called a *repository*, into which you *commit* versions of your work that represent logical pauses, like save points in a video game. Every commit includes helpful metadata like the name and email address of the person who made it, so you can pinpoint whom to praise (or blame) for a particular change. These commits are organized into *branches*, each representing an evolutionary track in your project's history, with one branch—the trunk, or master branch—representing the official, primary version. Having built up a history of past commits, it's easy to retrieve any previously committed version of your project, roll back changes, or compare two or more versions to aid in debugging.

In order to save changes to your repository, there needs to be one version of the project that you can safely make changes to. Version control systems like Git usually call this the *working copy*. Its job is to act as a scratch pad for any changes you may want to make to the project; you'll eventually commit these changes to the repository as an official, saved version. From our perspective, a working copy is usually easy to spot—it's the copy of the project that appears on our hard drives, as regular files.

Version control can seem laborious, because in a regular desktop workflow, we're expected to save changes to our work twice: once to the working copy, then *also* to the repository. As a young web developer starting out, I found this annoying enough to avoid version control altogether. Eventually, though, I came to appreciate the benefits of having every significant version of my projects stored, annotated, and neatly organized in a secure location. It also helped me to think of commits as *significant* changes, as opposed to the hundreds of little changes I might save in a given hour. The extra steps involved in committing—the brief pause from coding, having to write a descriptive message, occasionally having to stop and address conflicts between my version and someone else's—have ultimately helped me develop a more thoughtful and judicious way of working.

Although adopting a basic version control practice is a little extra work, it's not *hard* work. But like anything we do to stay organized, version control works best when it's practiced consistently, so it can become what *Getting Things Done* author David Allen calls a "trusted system." On one hand, once you've committed a version of your work to a repository, you should be able to trust that when you look for that version later, you'll be able to find it in exactly the same state as when you committed it. (As we'll see, Git has that part well covered.) But you should also be able to trust that the version you're looking for was committed to begin with, which means committing *yourself* to committing your changes at regular intervals as you work.

COMPLEX PROJECTS

Versions of single files, like Photoshop documents, are easy to manage: each represents a complete copy of the project as it existed at a certain point in its history, and (if you're using numbers or dates to identify versions) it's easy to tell when that point was by just scanning the list of versioned file names. But while some things we work on are neatly encapsulated in single files, others—like websites and apps—consist of entire directories of source files. How can we apply version control to projects like that?

This is where software version control systems like Git really earn their place in our toolbox—in managing more complex projects like websites or the source code for an app, or when coordinating changes from lots of collaborators.

The simplest version control method is the same one we'd use for a Photoshop file: make a copy of the whole project directory, appending dates or incremental version numbers as we go along. The directory named `our-website/versions/v12` is the twelfth revision to our project. Just as with a single file, every time we make a significant change to the website, we'll create a new numbered copy of the whole project: the one after v12 is v13, followed by v14, and so on. Here, the `versions` directory acts as our repository, and each numbered directory is a committed version. In simple cases, this works just fine. Indeed, before I got into using version control systems, that's how I managed web projects for clients.

Because a website consists of many different kinds of files, we need our working copy to be saved to the hard drive as its own directory: `our-website/working-copy`. The process for committing changes to this project is a bit more complicated than for our hypothetical Photoshop file, but only a bit: we make and test changes to the website in the `working-copy` directory; then, when we're ready to publish, we commit those changes by making a copy whose name includes the next version number in sequence, such as versions/v13.

But what happens when we try to share this version control system with other people?

These days, sharing files—such as our store of committed versions—is the easy part. Your repository can be a shared folder on a service like Dropbox or Google Drive, or a networked hard drive or file server in your office. Inviting a new collaborator to your project can be as simple as granting them access to that folder. Everyone who has access to the shared folder can potentially commit changes.

Where things get tricky is not simply in syncing changes between teammates, but in *coordinating* those changes so that your version control system remains trustworthy and viable. This system of numbered, versioned directories and working copies may seem straightforward, but you can never assume that two people will interpret or follow simple rules the same way. To sustain a trustworthy system, it's essential that the rules *always* be followed in *exactly* the same way. If you work on a small team and are thinking, "Come on, it's not that scary," imagine having to explain, let alone implement, a system like this within a large corporation, or an open-source project like Linux with *thousands* of contributors. This isn't just a random example: Git was invented by Linus Torvalds precisely to meet the demanding needs of the Linux project, after a licensing dispute about the commercial version control tool they were using (http://bkaprt.com/gfh/01-01/).

For the sake of argument, though, let's suppose that everyone on your team understands the rules, and that any member of the team can reliably commit a new version. Here's where things get *really* hairy: what happens when two people want to commit new versions *at the same time*?

The difficulty here doesn't involve following the rules so much as communication. Once we reach the point where two collaborators might need to work with the same files at the same time, we run the risk of one person's changes overwriting—or, to use my preferred technical term, "clobbering"—another's person's work.

Imagine you and I are collaborating on a website. You're exploring what it would look like if we changed all the links on the site from red to green, while I am considering changing the link color to blue. Under the rules of our version control system, I should make and save my changes to the copy of the stylesheet file in the directory. Unfortunately, the rules also say you should make *your* change in the exact same place. Following this method, whether the links are going to be green or red in the next numbered version of the project depends entirely on which of us made our change last. If I saved my file later than you saved your file, my changes win.

The only way to avoid this clobbering is for individuals interested in submitting changes around the same time—you and me, in this case—to work together to make sure that either our changes don't conflict or that our version control system somehow automatically reconciles or rejects conflicting changes.

Some primitive version control systems solve this problem by requiring people to "check out" a file, like a book from the library. Checking a file out marks it as uneditable by anyone else until the person who checked it out has both finished editing *and* explicitly checked it back in. This addresses the risk of unintended file-clobbering (by making accidental overwrites impossible), but it also creates a new problem: if I'm the one with the homepage file checked out, you're stuck waiting on me to finish before you can do your work.

DISTRIBUTED COLLABORATION

Instead of having one working copy shared by everyone, we can require all team members to have their own working copies stored on their own computers. In theory, at least, that allows each of us to work independently until it's time to save a new official version.

There's still a small risk of two people trying to commit versions at the same time, but that happens less frequently than just saving changes while working, and we can coordinate those kinds of changes easily—"Hey, I'm going to push version

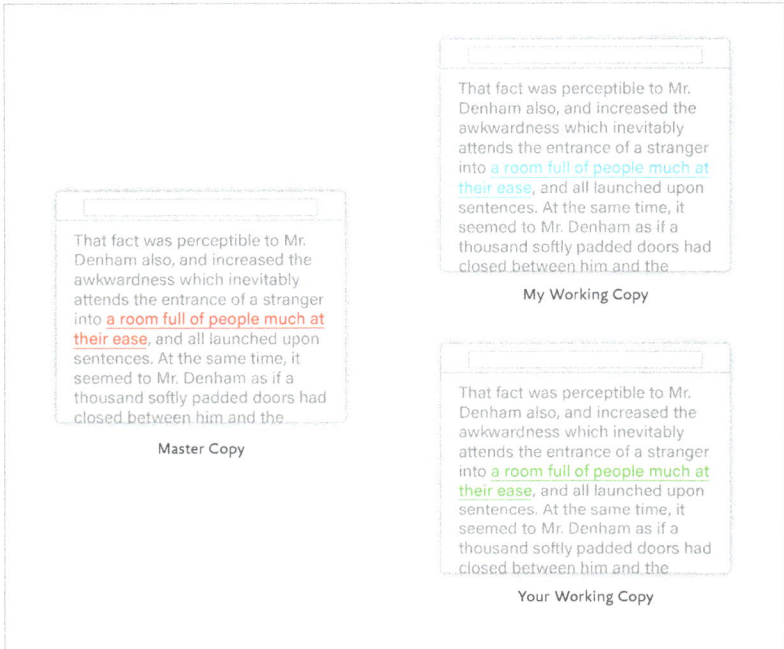

That fact was perceptible to Mr. Denham also, and increased the awkwardness which inevitably attends the entrance of a stranger into a room full of people much at their ease, and all launched upon sentences. At the same time, it seemed to Mr. Denham as if a thousand softly padded doors had closed between him and the

My Working Copy

That fact was perceptible to Mr. Denham also, and increased the awkwardness which inevitably attends the entrance of a stranger into a room full of people much at their ease, and all launched upon sentences. At the same time, it seemed to Mr. Denham as if a thousand softly padded doors had closed between him and the

Master Copy

That fact was perceptible to Mr. Denham also, and increased the awkwardness which inevitably attends the entrance of a stranger into a room full of people much at their ease, and all launched upon sentences. At the same time, it seemed to Mr. Denham as if a thousand softly padded doors had closed between him and the

Your Working Copy

FIG 1.1: Three different copies of this web page have different link colors. How can we know which one is the correct one?

34 of the website, everyone cool with that?"—via email or a tool like Slack.

With this system, we haven't done anything to try to merge together different versions, with different changes, into a cohesive whole. Rather, we're just making named copies of folders and assuming that our working copy is a trustworthy, canonical source for the next version. The next big problem to solve is what happens when that stops being true, and our working copies drift out of sync.

In the diagram in **FIG 1.1**, your version of the site now has green hyperlinks, while my copy has blue links, and the latest official version (v12) of the website has red links. Both working copies are newer than the shared master copy, but beyond that, how can we know what color the links should be in the next official version?

More importantly, how can we know what *else* might have changed? What if, in addition to the changed link color, your copy of the website includes a lovely new page explaining the company's mission that isn't in my copy, and mine has a fix for an annoying JavaScript bug that isn't in your copy?

This is where our homegrown version control system completely breaks down. It's not that we can't work together to combine our two working copies into a single version; it's that doing so takes up valuable time we'd much rather spend writing, designing, coding, making coffee, or browsing Tumblr for cat GIFs. Auditing our work files for conflicting changes is no fun; more importantly, it doesn't scale. As we add more files, more collaborators, and more changes, the risk of accidentally introducing major problems increases.

Remember that version control is more than just the versions: it's the rules and processes for *managing* versions. Once you have a lot of files, a lot of collaborators, or both, it can be exhausting to enforce those rules without a referee—and only by enforcing the rules consistently can you trust your version control system enough to get any value out of it.

Fortunately, all of these things—enforcing rules, keeping track of versions in a repository of past work, shuttling changes back and forth between the repository and your working copy, even merging together two directories and policing conflicts— are things that computers can do a lot faster and better than we can. By learning and adopting an *automated* version control system like Git, we can keep our work neatly organized and our changes safely coordinated with one another, all without a lot of effort—that is, once we adopt and learn how to use such a system.

HUMANS, MEET GIT

Git does for version control what web standards have done for our code or, for that matter, what Microsoft Word has done for word-processing documents. Because Git is so ubiquitous, once you know it, you can send code almost anywhere. Git excels at synchronizing changes between different computers, whether servers like GitHub or your colleagues' laptops. Far-flung members of your team can use Git to combine efforts on a project, pushing changes to a central hub where collaborators can pull down their own copies or review work in progress, and then use Git to push changes to a web server for deployment.

Git's way of staging changes and managing branches gives you unparalleled control and flexibility over how changes to your projects are committed and organized. These attributes make Git perfect for projects like the Linux operating system kernel (http://bkaprt.com/gfh/01-02/), with thousands of contributors and *hundreds of thousands* of commits. But Git also scales down beautifully for smaller projects and teams. Whether you're looking to add version control to your personal site or share code with your whole company, learning Git gives you a seat at a very awesome table.

Note, though, that Git isn't the only tool out there for automating version control or syncing files between collaborators. People sometimes rely on simpler file-syncing services, such as Dropbox, which offer shared folders and the ability to view and restore old versions of a given file. If most of your work involves single files—Word documents, spreadsheets, PSDs—and something like a shared Dropbox folder is working for you, you may not need Git.

HOW GIT WORKS

Git keeps your project in a local repository (usually a hidden folder on your hard drive). This is an important distinction between Git and older version-control systems like Subversion or the Dropbox-based versioning scenario mentioned earlier. These server-based processes are *centralized*, in the sense that the only place you can get at your whole history of prior versions is a shared, remote space, and only your working copy is accessible offline.

Git is a *decentralized* version control system. Both your working copy and a complete copy of the entire history of the project reside on your machine, the server, and every other computer that hosts a copy of the project. By default, Git's hidden repository folders live inside a visible working copy folder. If you browse that directory, you'll see only the files and folders you expect to see in a working copy of your project. This working folder is where you'll make your changes.

Whenever you're ready, you can easily move changes into the safe, stable repository by making a commit. In our semi-manual process, we "committed" a new version by making a copy of our working copy, naming it with the next sequential version number. Committing changes in Git is, conceptually at least, very similar. For each commit, Git records the precise state of our files as they are right now in the repository for later access and retrieval. Unlike in our manual example, where we had the annoying (and potentially risky) responsibility of making sure new versions were copied into place correctly without clobbering anyone else's efforts, Git automates all of that busywork. Even better, Git copies new versions incrementally, making references to existing copies of files that haven't changed to conserve disk space.

Git not only takes care of safely copying data back and forth between the working copy and the repository (and between the local and server repositories), but it also provides a robust system for referring to different versions of the project. One of the small costs of establishing a manual version control practice is needing to decide, and then communicate to your teammates, the correct way to identify single versions. Should you use a number (like v12), or a date stamp (2014-07-28), or something else? Git allows you to assign your own names or numbers to versions if you need to, but it also gives you a reliable, unique identifier for every single commit. If you don't need to assign custom names or numbers to versions, you can just sit back and rely on Git to do that.

Finally, Git also offers powerful tools for safely merging changes between different versions of a project—not just between different collaborators, but also between multiple variations of the project on one person's computer.

THE CHALLENGE OF GIT

Version control can be challenging for newcomers not (just) because it makes things complicated, but because change is legitimately complex. Using a tool like Git forces you to question your own assumptions about how change works.

For example, one of the things version control demands of us is a nuanced understanding of *state*. As humans working in a virtual space, we're used to applying physical metaphors as handy cognitive shortcuts for understanding digital things. Let's go back to our scenario of changing the link colors on a web page. Before our minds were trained in the philosophy of versions, we thought we had a file (like it was a physical object that just happened to be on the wrong side of a computer screen), and we were *changing* it. The CSS file remained constant, but the link color changed.

In fact, from the computer's (and Git's) perspective, there are at least *three* files: the saved copy from before we made the change (with blue links), the working copy where we replaced the line that controls link color, and then (finally) the new saved copy that replaces the old one.

But nothing about the mechanics of how this one line is updated changes the fact that styles.css appears to be one file *to us*. Semantically, viewing the pile of bits named styles.css as a single thing that changes is very valuable, because it helps us understand where to find our data. Having to spend too much time concerning ourselves with the difference between the versions of this file is annoying: it's better if we can rely on the name to tell us what file this is, and have some other way of understanding how it has evolved.

It's more accurate to say that, rather than three different files, we're talking about *the same file* in three different *states*. It's the same file because even though its contents may change, its name stays the same; *logically*, therefore, it's the same file.

One potentially confusing difference between our numbered file/directory names example and a true version control system like Git is that there is no giant folder full of old versions to look at. As Git users, we're expected to know that behind each *logical copy* of a file in our working tree, Git is safely storing all the old versions of the file, in each of its previous states.

We can comfortably understand a system where two files or directories are copies of each other, where one is a little newer or more evolved than the previous one, because there's an obvious real-world equivalent: manuscripts have second drafts, books have second printings, and so on. The rudimentary version control method I described earlier was relatively easy to understand because we were simply moving files around on a computer, something we've all done many times before. This new model feels less like writing drafts and more like time travel. It kind of *is* like time travel—and as anyone who has seen *Back to the Future Part II* can attest, time travel is complicated business.

For files saved on our hard disks, our apps and operating systems do a fantastic job of hiding—abstracting—all of that complexity from us. Instead of a flurry of versions moving back and forth between hard disk and memory, we just see an icon. Sometimes its contents change, and its "last saved" time is incremented accordingly, but visually and semantically it acts like the same file the whole time.

Not only does Git do a poor job of hiding that kind of complexity, it barely even tries. Git suffers from what I like to call an excess of simplicity.

Unlike many of the tools we use every day, Git doesn't do much to map the things it does to familiar metaphors or symbols, the way OS X maps deleting a file to the act of dragging it into the trash. Git's design assumes that you not only know how version control systems work, but specifically how *Git* works. You're meant to interact with Git on its own terms.

Git has more than 100 command-line functions, every single one of which has a specific job, with specific inputs and expected results. There is no single "Save new version" command in Git. Instead, to make a new commit—which is sort of, but not exactly, the same thing as saving a new version—you need to perform two or three different actions. Each of those actions has a legitimate purpose in Git-space, but none of them maps to something you'd logically do to a file or document in the real world.

But Git is also one of the most matter-of-fact programs around. It never does more than you tell it to do (though it can be easy to accidentally tell Git to do more than you wanted it to do). On one hand, this means that we might need to speak to it in more laborious, stilted terms than we're used to, which is itself an almost radical notion in an age when software can recommend a movie or summon you a taxi. On the other hand, the fact that the scope of a given command is limited means that there's also a limit to how much damage you can do at any one time. If you *do* get into Git's version of trouble—like if Git can't easily reconcile conflicting changes, or if it's uncertain about where to commit your work—there is *always* a command that will get you out of it, often with whatever work you were trying to save still intact.

As we've seen, a good argument can be made for even small teams to use version control. But the internet has made it easier than ever for people on opposite sides of the globe to work together on all kinds of projects. The open source movement has taken that even further by creating opportunities for thousands of strangers to contribute changes to projects seen and used by millions—collaboration on a wildly unprecedented scale. When you're trying to accept contributions from a community of thousands, version control becomes an absolute necessity.

So although version control may have started out as a form of insurance against mistakes, tools like Git have helped transform it into something much more compelling. As both a practice and a set of tools, version control offers us a common framework for collaborating on and sharing all kinds of work with anyone, anywhere—not to mention a new way of understanding and managing change. Thinking and working in versions not only helps us understand how projects evolve over time, but also gives us more say in how that evolution happens.

Now, how can we begin to incorporate thinking in versions into our workflow?

2 BASICS

IF YOU'RE JUST STARTING TO LEARN GIT, I recommend sticking to the command line, at least at first. Git's command-line interface is its native tongue. Typing commands and seeing the responses Git gives back is a great way to learn about how Git actually works, which will pay off when you inevitably run into a confusing situation down the road. The command line is also consistent across the various platforms Git runs on. If you know how to interact with Git via commands, you'll be able to use Git no matter what kind of computer you're on.

The command-line interface gives you the fullest access to all of Git's powers, but there are also graphical Git apps out there, some of which are very good. And although I recommend starting out with the Terminal, you don't have to choose one over the other. You can in fact use the Terminal and a Git app side by side, making commits and pushing changes in the app, where it may be easier (or more your style), and relying on the command line for everything else.

For now, let's start with something relatively easy: getting Git on to your computer.

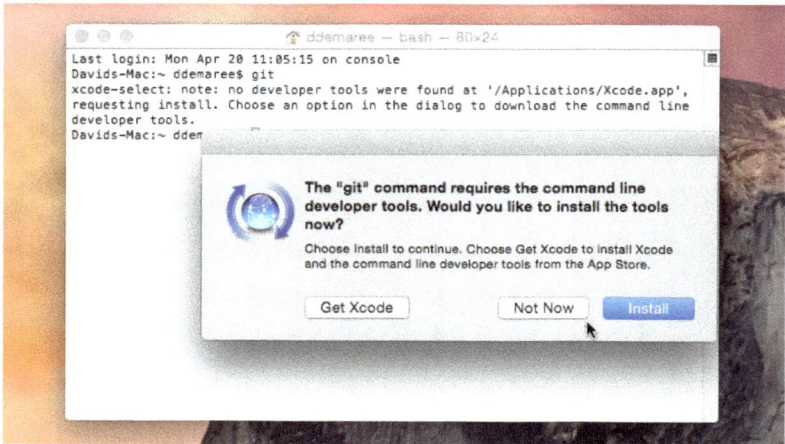

FIG 2.1: If you open up a Terminal window and type `git`, your Mac will offer to automatically install an Apple-maintained software package that includes Git.

INSTALLING AND RUNNING GIT ON A MAC

Starting with version 10.9 (Mavericks), released in 2013, OS X automatically downloads and installs command-line tools like Git the first time you try to use them. If you simply open up the Terminal app and type a Git command, your Mac will prompt you to install a package called Apple Command Line Tools, including Git and several other utilities. After installation, OS X will automatically download and install updates to Git via the Mac App Store app.

INSTALLING AND RUNNING GIT ON WINDOWS

The Git development team maintains an easy-to-use installer package for Windows that you can download from the official Git website (http://git-scm.org/). The install wizard will ask you a bunch of questions about how you want to configure and use Git; if you're unsure how to answer any of them, just go with the default settings.

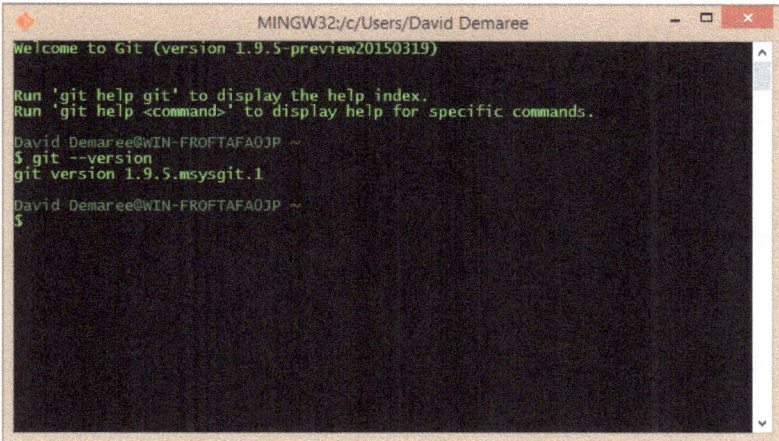

FIG 2.2: The Git Bash application that comes as part of Git distributions for Windows. It emulates a complete Unix-style shell, with the same commands you'd find on Linux or OS X.

If you're comfortable with either of Windows' two standard command-line environments, Command Prompt or Power-Shell, the Git installer will give you the option of configuring Git to work with those. By default, however, the Git installer provides its own terminal application, called Git Bash, which emulates (that is, works similarly to) a Unix-based system such as OS X or Linux, with support for not just Git but all of the commands we'll be using in this chapter and throughout the book. If you're accustomed to using Git on one of those platforms, or if you want the most consistent command-line experience across different computers, Git Bash is an awesome tool.

Although Windows and Linux/OS X both support the same encodings for plain text files (ASCII and Unicode), they use different line endings, that is, different characters for denoting line breaks. Many Windows or cross-platform text editors, including Atom and Sublime Text, already do a good job of handling line endings correctly, but Git may not always know which format it should use.

Git for Windows helps us out here by offering to automatically convert Windows line endings to Unix ones when you commit, and vice versa when you check out. Even if all of your teammates and servers run Windows today, it's usually best and most future-proof to commit Unix line endings to ensure that your files can be checked out safely on non-Windows systems.

One more small difference: while Windows paths use backslashes and drive letters, as in `C:\Users\David\myproject`, Unix paths use forward slashes. An equivalent path on my Mac would be `/Users/David/myproject` (with no leading drive letter, since Unix systems like OS X don't use them). Here, too, Git Bash helps us out: it automatically swaps slashes, presenting that Windows path as `/c/Users/David/myproject`. (The leading `/c/` indicates the `C:/` drive, following Unix convention of referring to mounted disks as directories at the topmost, or root, level of your computer's file system.)

COMMAND LINE BASICS

If this is your first time interacting with a command line, entering weird bits of jargon after a prompt may seem scary unless you're a programmer, sysadmin, or computer scientist. It's an understandable fear: the command line *is* a little scary. Your computer will let you do things from the command line that it would rightly stop you from doing from the Finder or Control Panel—like delete your operating system, not to mention your entire hard drive. Keep in mind, however, that although the command line is powerful, on another level it's also a little stupid: it will only do what you tell it to do, and most commands are set up to perform a single, simple task.

Before we dive into the specifics of using Git, I'll go over how command-line interfaces work, and how to read the examples that appear throughout the book.

What the command line is for

The command line is a holdover from an era before fully graphical operating systems like Windows and OS X were powerful enough to govern every aspect of a computer. If you're as old as I am, you may remember having to bust out of the Windows environment back to MS-DOS for one reason or another. You may also remember watching *Back to the Future Part II* as a newly released VHS tape.

The command-line interpreter itself is called a *shell*. Like so many other computing words, it's a metaphor. Just as some animals, like turtles, have shells that conceal their fragile or dangerous bits, with openings left for the creatures to move around and interact with their environments, a software shell covers and protects the computer—and you, the user—from commands that might cause it to behave erratically or stop working altogether. Although we're mostly talking about command-line shells here, the visual interfaces we're familiar with today are *also* shells—they're just shells made of images instead of text.

The opening in our metaphorical shell is the command-line prompt, a string of characters signifying that the computer is ready to receive a command from you. Most of the prompts in this book will look like this:

```
$: _
```

The underscore (_) character here represents the cursor; the `$:` is the prompt itself, which signals that the shell is waiting for you to ask it something.

Although shells are our primary means of interacting with some kinds of systems, they're really just programs. The default shell on OS X and many Linux systems, and the one included in the Git package for Windows, is called *bash*. Bash is the most widely used command shell for Unix-like computers, and although it's hardly the only one, it's the one that, for the purposes of this book, I'll assume you're using.

You've probably seen movies or TV shows where hackers in dark sunglasses data-mine mainframes and infiltrate Gibsons, all by typing what appears to be gibberish into computer screens full of more gibberish. The one thing these movies often get right is how a command prompt works. First, you type the thing you want the computer to do in a language it understands:

```
$: whoami_
```

Then you hit Enter to submit the command. In this case, we've typed the name of another program we want the shell to run for us. This one, whoami, is very basic: it just tells us what username we're logged in as. Once it has finished running, which is typically right away, the program will output (or "print") a response into the Terminal window below the line where we entered the command, like so:

```
$: whoami
david
```

After that, it'll print the prompt characters again, ready for another question:

```
$: _
```

Command-line programs often lack ceremony. They'll print out the information you asked for but won't try very hard to make it pretty or put it into context. Where you might expect to see a label or some contextual information—"Your user name is: david"—the whoami command just tells it to you straight. You asked whoami, it answers david, and then gives you a fresh prompt to ask another question. Many commands (both included in the operating system and included with Git) return no response at all.

Although many commands, including several of Git's, can be almost too chatty with their responses, whoami is terse almost to the point of rudeness. That's not a sign that anything is wrong. On the contrary: the lack of response indicates that the system and shell trust you to know what you're doing. A program finishing (or "exiting") with no output almost always means that the job you asked of it was done with no errors.

For example, here's the mkdir command, which creates a new directory:

```
$: mkdir javascripts
$:
```

In case it's not clear what just happened: mkdir didn't provide any feedback at all, which could lead you to think that the command didn't work. But watch what happens if we try it again:

```
$: mkdir javascripts
mkdir: javascripts: File exists
$:
```

In this case, the command fails for a simple reason: a directory called javascripts can't be created because it already exists.

Command-line navigation

The Terminal, like a single Finder window on a Mac, is always looking at a single *active* directory or folder. This is called the *working directory*. You can find your current location as an absolute path using the pwd ("print working directory") command:

```
$: pwd_
/Users/david
```

We can do that using the cd ("change directory") command, which sets a given path as our new working directory—the command-line equivalent of double-clicking on the folder's icon. This will often be the first command you run in a new Terminal session, to navigate from your shell's default starting point (typically your home directory) to the directory containing your Git project.

```
$: cd Projects/our-website
$: pwd
/Users/david/Projects/our-website
```

Here we've typed in the path to our project directory (Projects/our-website), relative to my home directory, where we started this Terminal session. Relative paths are distinguished from absolute ones by the absence of an initial slash

(/), which here represents the root—or topmost—directory of your local file system.

One handy tip to remember if this is your first experience with the Terminal: you can use the tilde character (~) in paths as an alias for your home directory. The absolute paths /Users/david/Projects/our-website and ~/Projects/our-website are equivalent, and you can use either one to jump to your project directory from anywhere else on your hard drive.

To see a list of all the files and directories inside the current directory, use the ls ("list") command:

```
$: ls
css           index.html
```

This particular directory has two things in it: a web page (index.html) and a subdirectory containing our stylesheets (css). Unlike in the Finder, there are no icons to differentiate the different kinds of things we're looking at, like whether css is a directory or a file. But ls gives us a couple of options for requesting more information. For instance, if you tack on the -F option, ls will show trailing slashes to help distinguish directories from regular files:

```
$: ls -F
css/          index.html
```

To look inside a directory without navigating into it, we can pass the name of the directory to the ls command as an argument:

```
$: ls css
styles.css
```

Of course, if we need to, we can navigate into a subdirectory of our project using cd.

```
$: cd css
$: ls
styles.css
```

As on the web, the double-dot (..) symbol represents the current directory's parent, and, as you might expect, it works from a shell prompt.

```
$: ls ..
css             index.html

$: cd ..
$: pwd
/Users/david/Projects/our-website
```

It's helpful to remember that the command line and graphical tools like the Finder essentially do the same thing. They both look at the same files on the same hard drive, which means that you can use the Finder (or an app's Save... dialog) to save files or create directories if you prefer a graphical interface. If you're handier with the command line, or don't like switching back and forth from Git, it may be faster to do almost everything in the Terminal. On the other hand, depending on your experience and preferences, you may find that you're able to work more quickly and efficiently using the Terminal and Finder in tandem than by using either one alone.

The prompt

To keep things simple, the command prompts used for the remainder of the book will look like this:

```
(master) $:
```

The current Git branch, in this case master, is listed here in parentheses. Later on we'll explore other branches, but master is the branch we'll be working on in this chapter and the next one.

In our examples, an asterisk after the branch name will indicate uncommitted changes in our working copy, like so:

```
(master *) $:
```

Every operating system and shell has its own default prompt format, and if you don't like the default, you can customize the prompt to look any way you want. It's very likely that your computer's prompts and the ones in this book will not look alike.

I bring this up mainly to reassure you that it's fine if the examples in this book and the prompts in your terminal app don't look exactly the same. The Git commands and their output are what matter most, and those should be consistent no matter what your prompt looks like.

TALKING TO GIT

Now that you know how to navigate the Terminal, it's time to start really interacting with Git by sending it commands, which all look more or less like this:

```
(master) $: git commandname parameter1 parameter2 »
  --option
```

The command name (commandname in the example) is one of *over 100* individual functions that Git can perform. Behind the scenes, each of these commands is a separate program responsible for its own specific job.

Though some Git functions work with just a command name (like git status), most require some parameters to know how to do their jobs, similar to passing input to functions in a programming language like JavaScript or Ruby.

You can read a lot of Git commands as a kind of sentence: *Git, please do a thing to this other thing*. For example, the command git checkout master essentially means: "Git, please check out the branch named master."

Options are special parameters that are denoted by at least one leading dash character. These are rarely required, and usually change something about the default way Git handles a particular task. Many options have both a long form, like --global, and a shortcut form, like -g. There are also options that take values, like `git commit --message="hello world"`. As we go along I'll call out important options, what they're used for, and the kinds of values they take.

CONFIGURING GIT

Git is a complex beast with hundreds of configuration options, but there are two that it absolutely needs in order to function: your name and email address. Git adds an `Author` attribute to every commit you make that includes both your name and email address, so that your collaborators on a project can know who made a given change. The name you enter will be used to identify you in change logs and any other place where Git shows who made a particular change, while your email address not only tells people how to reach you, but also tells a hosted service like GitHub who you are on their service.

So let's tell Git who you are, using the `git config` command. Unlike most Git commands, which only work inside of a Git project, these can be run from any directory.

Enter each of these lines at a command prompt, filling in your own name and email address:

```
$: git config --global user.name "David Demaree"
$: git config --global user.email "david@demaree.me"
```

Here, we're telling Git that a particular configuration property (`user.name`) should be set to the value (`David Demaree`) we've provided.

The `--global` option tells Git to set these values as a default for all projects on this computer. You can, if you want to, use the `git config` command to set configuration options like `user.name` within specific projects by just omitting `--global`. But we're setting these globally for now because Git requires them to be set *somewhere*, and this way you won't have to do it every time you start a new project.

A brief note here about privacy, because sharing personal information such as this can be a sensitive topic. Git will use the name and email address you give it to provide attribution for any commits you make. For local repositories, or commits that haven't been pushed to any server, this information will reside only on your computer. But commits are meant to be shared, so be aware that commits you share with others (whether it's within your team or on an open source project) will include this information. What's more, as you'll see, commits are not really meant to be changed after the fact.

I mention this to alert you, not to scare you. While Git requires you to provide a name and email address in order to attribute commits, it doesn't know or care whether the values you enter are your real name or email address. In fact, pairs of programmers working on the same problem on the same computer will often change their Git setup to claim joint credit for any commits they make.

When you fill out these values, you are free—and expected—to provide only as much information about yourself as you feel comfortable providing.

STARTING A NEW PROJECT

Once you've adopted Git as your version control system of choice, creating a new Git database using the `git init` command will usually be one of the first things you do. But it will almost never be the *very* first thing, because most Git repositories are designed to live inside a folder on your hard drive—the working directory—alongside your project files.

More simply: in order to track changes within a project folder, first you need to have a project folder. This may be something you're already working on and are adding Git to, or it may be something new, for which you want to use Git from the get-go.

For the purposes of this chapter, we'll start developing a new website. First, let's create a directory to work in. Though we could do this from the Finder or Windows Explorer, in this case we'll do it from the command line.

```
$: mkdir our-website
```

This creates a new folder called our-website inside of the current directory. Next, let's switch into our new directory using the cd command:

```
$: cd our-website
```

Now let's initialize a new Git repository within this new project folder, using the git init command.

```
$: git init
Initialized empty Git repository in /Users/david/ »
  work/our-website/.git/
(master) $:
```

Boom! We now have a fresh, new, empty Git project on our computer. It's empty both in the sense that it has no files yet—remember, we just created this directory for the first time—and that it has no commits.

CLONING AN EXISTING PROJECT

If you're not the person responsible for initiating your project, it's more likely that your first step upon joining a Git project will be to pull down a copy of the repository stored on a server somewhere. This is called *cloning*, meaning that what gets saved to your computer is a replica of everything that was stored on the server.

Cloning a repo is a sequential process that Git helpfully wraps in a single command: git clone. A sort of *macro*, git clone is a single, convenient command that performs several related commands at once: it creates a new working directory (named after the repository on the server by default); initializes a new Git repository; adds a remote called origin; and pulls changes from the remote.

```
$: git clone https://gitforhumans.info/ »
  our-website.git
Cloning into our-website...
remote: Counting objects: 11, done.
remote: Compressing objects: 100% (7/7), done.
remote: Total 11 (delta 1), reused 11 (delta 1)
Unpacking objects: 100% (11/11), done.
Checking connectivity... done
$: cd our-website
(master) $:
```

Following the default of automatically naming the working directory after the repository name is almost always the simplest way to proceed. But if you *do* want to give your directory a different name—let's say you want to prepend a client's name—you can simply pass a different folder name into git clone as an argument. Here, instead of naming this copy of our client's project our-website, let's call it clientco-website:

```
$: cd ~/Work
$: git clone https://clientco.co/our-website.git »
  clientco-website
Cloning into 'gfh-website'...
remote: Counting objects: 11, done.
remote: Compressing objects: 100% (7/7), done.
remote: Total 11 (delta 1), reused 11 (delta 1)
Unpacking objects: 100% (11/11), done.
Checking connectivity... done
$: cd clientco-website
(master) $:
```

Either way, once you've cloned the remote repo, everything about Git works the same as if you had created the project yourself. What's more, since Git is generally concerned only about what's inside your project, not the folder containing it, no matter what name you give when cloning the project, you can safely rename the folder anytime you like.

The clone we've just created comes with the full history of this repository, including every change shared by every other copy. This notion of being able to clone not just the working state, but also the entire history of the project, will come in handy later.

GETTING READY TO COMMIT

Git expects every change you make within a directory under its care to be recorded and stored in the repository for safekeeping—even the change that takes a project from nothingness into being. The phrase *commit or it didn't happen* sounds like a joke, but it's also literally true. As a practical matter, Git is more concerned with managing your commits than with the files whose changes you're using commits to track. It's not exactly that Git is indifferent to the contents of your files; it's just that its model for organizing and managing your work is oriented around commits.

Commits are a type of Git data called an *object*. Internally, every piece of information Git knows about—the contents of files, the structure of folders, and, most importantly, the commits that mark the others as versions of a project—is stored as an object, and each particular object has a unique name derived from its contents. Really, the name—or identifier—of an object is less a name than a short, machine-readable fingerprint that reliably distinguishes objects from one another.

Commits are the only kind of object you'll work with on a daily basis. Semantically, each commit represents a complete snapshot of the state of your project at a given moment in time; its unique identifier serves to distinguish that state from the way the files in your project looked at any other moment in time.

Git proceeds by addition. Even though files in your project can be created, deleted, or changed, the commits tracking those changes are always *added*. When you remove a file, you're adding a commit. If you change a line of text or code, or even change a file's name, you're changing the state of your project, and you'll add a commit to mark that change and propagate it to the rest of your team.

This is part of the beauty of Git's design: items in its database are lossless, *immutable*: they can never truly be changed; only added to. Git is a system of accumulation. It accumulates every change you tell it about, so that you can go back and explore that history later on.

When you commit a change to your work, some really cool stuff happens behind the scenes, which we'll look at in more detail in the next chapter. For now, though, let's make a commit to see how that works.

UNDERSTANDING YOUR STATUS

So, we've created a directory to hold our website project. Next, let's add a new file to serve as our homepage. Start a new document in your favorite HTML editor, and add this to it:

```
<!DOCTYPE html>
  <html>
    <head>
      <title>Our Website</title>
    </head>
    <body>
      <h1>Our Website</h1>
    </body>
  </html>
```

Done. Save the file to your project directory as index.html.

Before we move on, let's ask Git what it knows about the state of our project using the git status command.

```
(master *) $: git status
# On branch master
#
# Initial commit
#
# Untracked files:
#   (use "git add <file>..." to include in what will
   be committed)
#
# index.html
nothing added to commit but untracked files present
(use "git add" to track)
```

There's a lot of interesting information here. First, we learn that we're on our project's master branch; we'll talk more about that in Chapter 4. For now, rest assured that we're in the right place within the ridiculous multiverse that is Git. Because this is a new, empty repository, our next commit will be the *initial* commit—that is, the very first one on our timeline. Finally, and most importantly, we have a list of "untracked" files, which includes our index.html file.

ADDING FILES TO GIT

Before you can commit a file, it must be *tracked*; before a file can be tracked, you have to add it to Git's database. That confuses a lot of people who are new to Git, because aren't adding a file and committing it the same thing?

No, they are not! A commit records changes to files in Git's database—to say that, for instance, a particular file went from version A to version B (or, in the case of the initial commit we're working on, to define what version A *is*). Logically, before Git can know how A has changed to become B, it has to know about versions A or B individually. For the sake of—please don't laugh—simplicity, it's normal for us humans to treat commits as a shorthand to represent specific versions of files and folders. But the commits themselves are just references, similar to the way a street address references a house.

When we add a file, we are building the house: the `git add` command makes a snapshot of the given file and saves it to the repository so that it can be referred to later in a commit.

This means that sometimes Git saves snapshots that will never be committed, and that's fine. These take up very little disk space, and from time to time, Git will do something called *garbage collection* wherein it finds objects that aren't referenced by any commit and deletes them. If that sounds harsh, think of it this way: you haven't committed to the versions of your work represented by these stray objects. If they were worth saving, Git assumes you would have committed them somewhere.

By now I'm sure you're like: "That's interesting and all, but now how do we get this first version of our file into Git's database so we can commit it?"

As the `git status` message suggested, we'll use the `git add` command.

To add the homepage file, type this:

```
(master *) $: git add index.html
```

Git generally won't give you any response to tell you anything happened, and once you've done this a few times you won't need one. But since this is your first commit, let's check `git status` again:

```
(master *) $: git status
# On branch master
#
# Initial commit
#
# Changes to be committed:
#   (use "git rm --cached <file>..." to unstage)
#
# new file:   index.html
#
```

Now we see that our file is no longer listed under Untracked files, but rather under Changes to be committed. We have now *staged* our file, and we're ready to commit.

```
(master *) $: git commit --message "Initial commit"
[master (root-commit) 600df9f] Initial commit
 1 file changed, 9 insertions(+)
 create mode 100644 index.html
(master) $:
```

Boom. You just committed your first file.

If you look closely, on the first line of the response, right before the commit message we gave, you'll see this commit's unique ID: 600df9f. Most commit IDs you'll see (and we'll see quite a few of them as the book goes on) will look a lot like this one.

The email and name we added earlier tell Git who you are, and the --message option tells your collaborators (or your future self) the nature of the change we just made. While we're here, let me save you some typing with a little trick. Many Git command options are available in a shorter (usually single-character) form, denoted by a single dash instead of double dashes. For example, the --message argument can also be typed as -m.

```
(master *) $: git commit -m "Initial commit"
```

With that, we've made our first step along the long road of the history of this project. Let's move on to our next change, and our second commit.

THE STAGING AREA

Before you can commit a new version of your files, that new version must be added to Git's database, something we do with the git add command. Another name for this is *staging*; the *staging area* is where these new versions live between when you update your working files and when you commit them.

Staging a file causes two things to happen behind the scenes. First, Git saves a snapshot of that file to its database, so that it can be referred to in your next commit. The nature of Git references is such that a file must already be in the repository for it to be referred to, and it must be referred to in order to commit it. Until a version of a file is staged, Git doesn't know how to refer to that version, and therefore can't commit it.

Git also starts a local draft of your next commit, with references to all of the files and directories contained therein—as it happens, including references to files that haven't even changed, copied over from the previous commit. Every commit is self-contained: it doesn't just reference the things that have changed; it references *everything* that makes up the state of your project at a given moment. Most of the time you won't need to know the mechanics of how this works, but I find that understanding what's going on helps me make better sense of the commit workflow.

Unlike most of the data in your Git repository, the staging area is not synced or shared with anyone else on your team—it lives only on your computer.

OUR SECOND COMMIT

This first commit may have seemed like it took a long time, what with me digressing to explain data stores and the semantic nature of Git objects, but as we go on you'll find that these three commands—status, add, and commit—will make up the bulk of your interaction with Git. You may use this basic commit workflow dozens of times a day and, barring any especially tricky situations, these three commands will be all you'll need.

To illustrate this, let's make our second commit, adding a basic CSS file that we'll link to from our home page. First, within your project directory, use the mkdir command to create a new subdirectory called "css". Then fire up your text editor, open a new file, and type:

```
body {
  font-family: 'source-sans-pro', Arial, sans-serif;
  font-size: 100%;
}
```

Call this file "styles.css" and add it to the css/ subdirectory. Next, let's link to this new stylesheet from our index.html file:

```
<!DOCTYPE html>
<html>
  <head>
    <title>Hello World</title>
    <link href="css/styles.css" type="text/css" »
      media="all">
  </head>
  <body>
    <h1>Hello World</h1>
  </body>
</html>
```

To recap what just happened in the file system: first we added a directory (css/) with one file in it (styles.css); then we changed a file that already existed. Now let's check git status:

```
(master *) $: git status
# On branch master
# Changes not staged for commit:
#   (use "git add <file>..." to update what will be
  committed)
#   (use "git checkout -- <file>..." to discard
  changes in working directory)
#
# modified:   index.html
#
# Untracked files:
#   (use "git add <file>..." to include in what will
  be committed)
#
# css/
no changes added to commit (use "git add" and/or
  "git commit -a")
```

Under the heading Changes not staged for commit, we see the HTML file Git already knows about (because we committed it earlier), which Git now (correctly) says has been modified. Below that, the Untracked files list is back; instead of the styles.css file we added, we see the css/ directory that contains it. This is Git's way of telling us that there's an entire subdirectory it doesn't know about.

We can stage both of these changes with a single `git add`. Here, we're going to list *both* our HTML file and the whole `css/` directory as arguments, separating them with a space to indicate that we want to stage both of the things in this list. In English, it's like we're saying, "Git, please add this and this."

```
(master *) $: git add index.html css/
```

Let's check `git status` again:

```
(master *) $: git status
# On branch master
# Changes to be committed:
#   (use "git reset HEAD <file>..." to unstage)
#
# new file:   css/styles.css
# modified:   index.html
#
```

Both files are staged and ready to commit, with Git now noting that index.html has been modified from a previous version, while styles.css (and the `css/` directory it lives in) are new in this commit. Let's commit.

```
(master *) $: git commit -m "Add stylesheet"
```

With this second commit, your project now has history. We'll explore Git's commit log more in the next chapter, but you can already see a timeline taking shape when we run `git log`.

The process of adding files and then committing them will cover a surprising amount of your version-control needs. And so far, everything makes (relative) sense. The need to stage files before committing them may seem a little strange but, as we'll see later, it can also be powerful. At any rate, right now it's just a small, extra hassle, not a fundamental change in the way we manage files.

But other common kinds of changes are less intuitive. We'll go over some of those next.

REMOVING FILES FROM GIT

When we delete a file in our working copy of a project, it follows that we should also be removing it from our repository. To be sure, Git's command for deleting files—git rm—does its best to act like it's simply deleting a file.

Having said that, let's recall two things we've already learned: that Git is a system of accumulation, and it only cares about changes in the context of a commit. This brings us to our first serious logical paradox in working with Git.

For those of you new to the command line, rm (short for *remove*) is the standard Unix file-deletion command. From a command-line prompt, typing rm path/to/my/file will delete the file at that path. git rm behaves in a very similar way, with one added benefit: in addition to deleting the file, it also stages a new commit where the file has been deleted. In other words, in order to remove a file, we have to *add* a commit.

That last statement may seem a little mind-boggling, so here's an example that illustrates how files are removed in Git.

Let's say that since the last time we worked with our web project, someone has added a robots.txt file telling search engines not to index anything on our site. (We'll go over how other people's changes get into our repos later. For now, imagine that time has passed and that our project has picked up changes.)

Now, though, we've changed our minds and have decided we actually *do* want to be indexed, so we need to remove the robots.txt file.

To do that, we'll use git rm:

```
(master) $: git rm robots.txt
rm 'robots.txt'
[master *] $:
```

If we run git status, we'll see that the file's deletion has been staged for inclusion in the next commit:

```
[master *] $: git status
# On branch master
# Your branch is ahead of 'origin/master' by 1
  commit.
#   (use "git push" to publish your local commits)
#
# Changes to be committed:
#   (use "git reset HEAD <file>..." to unstage)
#
# deleted:    robots.txt
#
```

Git's snapshots, upon which it bases commits, consist of your files' contents and the directory structures that contain them. When we delete this file using git rm, Git creates a new snapshot of the project *minus* robots.txt, and stages that version as the next one to be committed. Now let's commit it:

```
[master *] $: git commit -m "Remove robots.txt"
[master 983024f] Remove robots.txt
 1 file changed, 0 insertions(+), 0 deletions(-)
 delete mode 100644 robots.txt
```

Git vs. trash

Of course, you may be more accustomed to deleting a file by simply dragging its icon into the Trash (or Recycle Bin). Or if you use an all-in-one editing tool like Coda, maybe you delete files by using its built-in file manager. Even when you know Git has a "remove file" command, it's hard to overcome years of muscle memory when the delete command you've used for so long is *right there*.

I sympathize with this because even I, a programmer who has used command lines and Git for a really long time, almost always delete files the old-fashioned way (that is, whatever way is most convenient), and then tell Git about the deletion afterward.

As it happens, the command to do this is the exact same one we just used, git rm. If you delete a file yourself, git rm's job is limited to staging a new change that removes it from the Git index.

So, after deleting our robots.txt file the old-fashioned way, we can just run:

```
[master *] $: git rm robots.txt
rm 'robots.txt'
```

Git still gives us a response letting us know it ran the rm 'robots.txt' command to delete the file from our hard drive, even though it didn't need to. That's fine—the rm command does nothing if the file has already been deleted.

RENAMING FILES, OR: GIT'S ABSURD RELIANCE ON NAMES

Next, we want to rename our website's main stylesheet from styles.css to screen.css (to make space for a separate print.css we might add later). Muscle memory being what it is, we're liable to use whatever file-renaming command is most familiar. So let's say we've renamed this file in the OS X Finder, and now we check git status:

```
[master *] $: git status
# On branch master
# Your branch is ahead of 'origin/master' by 2
  commits.
#   (use "git push" to publish your local commits)
#
# Changes not staged for commit:
#   (use "git add/rm <file>..." to update what will
  be committed)
```

```
#    (use "git checkout -- <file>..." to discard
   changes in working directory)
#
# deleted:      css/styles.css
#
# Untracked files:
#    (use "git add <file>..." to include in what will
   be committed)
#
# css/screen.css
no changes added to commit (use "git add" and/or
   "git commit -a")
```

We now have an unstaged change deleting our stylesheet entirely. What? Where did it go?

If you look under Untracked files, you'll see it there—but Git is telling us that it thinks our renamed file is a totally new, untracked file. This probably seems crazy to a human, but from Git's extremely literal, name-based perspective on the world (and, more specifically, on your file system), it's all too logical.

Git is saying that the file it was tracking named "css/styles.css" is no longer present *under that name*. Meanwhile, it's also saying that it's not tracking a file called "css/screen.css" because we haven't asked it to track a file *by that name*.

Of course, *we* know that these are just two names for the same file. But Git doesn't know that, because Git relies on names to know whether a particular file is familiar to it or not. It may seem simple or logical to *us* that this was just a name change, but in order to avoid making a bad assumption about a change that could result in an incorrect commit, Git makes no assumptions when you change things via any method other than a git command.

Because it just seems so easy for Git to take care of this for us, this kind of thing might frustrate you when you first encounter it. But I view it as Git's simplicity at work. The job of Git is to track changes and commits. Period. Git can figure out that the deleted file styles.css and the untracked file screen.css have the *exact same contents*, but it has no idea what that *means*. It doesn't (can't, really) make the leap required to assume that the two paths are different names for the same file, because their being the same file is *semantic*—it's meaning that you have *ascribed* to it, not something intrinsic to the data.

For instance, what if you wanted to create a second copy of this file under a different name? Or what if you deleted styles. css by accident? Git must allow for any of these possibilities, no matter how silly they may seem to us.

But back to our git status: we're left with one missing file and one mysterious new file. *We* know they're the same file, but Git does not. Let's first try the git mv (short for "move") command, which is Git's typical renaming function. Here's what happens:

```
[master *] $: git mv css/styles.css css/screen.css
fatal: bad source, source=css/styles.css,
  destination=css/screen.css
```

Unlike git rm, which doesn't care if the deleted file has been deleted or not—and, as we saw, doesn't even need for the file to actually be deleted—git mv will only rename a file if it is also allowed to move or rename the files in the working copy. If we'd used git mv to do this initially, instead of the Finder, it would have worked flawlessly. But we didn't, so now we need to figure out how to stage and commit this change another way.

Git sees two uncommitted changes—a deletion (css/styles. css) and a new addition (css/screen.css)—and we need to address each one individually before we can commit.

First, we'll use git rm to stop tracking styles.css:

```
[master *] $: git rm css/styles.css
```

Then, we'll use git add to tell it to start tracking the file under its new name:

```
[master *] $: git add css/screen.css
```

Once we do both of those, we check our status:

```
[master *] $: git status
On branch master
Changes to be committed:
  (use "git reset HEAD <file>..." to unstage)

    renamed:    css/styles.css -> css/screen.css
    modified:   index.html
```

This is the *exact same response* we would have seen if we had used `git mv` to begin with. Deleting, moving, or renaming files with Git's built-in commands can save you some typing, but it isn't necessary. This is an example of a scenario you may encounter while using Git that seems like trouble, but really is just annoying. Depending on how often you rename files, this will either be an incentive to always do things Git's way, or else will make you feel comfortable doing things in a way you're familiar with, knowing that you can always explain yourself to Git later.

THE WHAMMY: git add --all

When in doubt—or running short of time—there's a nuclear option for quickly staging anything and everything that has changed in your local copy: `git add --all` (or `git add -A`, for short). The `--all` option is great for moments when you need to commit several changes at once. For example, we've started to add some JavaScript behavior to our web page, adding a new file (and directory) called `js/site.js` and linking to it from our HTML document. Here's our status:

```
[master *] $: git status
On branch master
Changes not staged for commit:
  (use "git add <file>..." to update what will be
  committed)
```

```
    (use "git checkout -- <file>..." to discard
    changes in working directory)

        modified:    index.html

Untracked files:
    (use "git add <file>..." to include in what will
    be committed)

        js/
```

The js directory (containing our script file, site.js) is new to the project, and thus shown as untracked. We also edited index.html to add a script tag linking to the new file; it's shown here as modified but unstaged.

Normally we would notify Git of these changes one by one, typing in the paths for each file and directory we need to stage. Instead, let's use git add --all, then check our status:

```
[master *] $: git add --all
[master *] $: git status
On branch master
Changes to be committed:
    (use "git reset HEAD <file>..." to unstage)

        modified:    index.html
        new file:    js/site.js
```

With that one command, everything we've done is ready to commit.

That was a fairly straightforward example, but we can make it more complicated. Before we commit, we decide to rename the directory containing our scripts, from js to the more descriptive scripts, using the mv ("move") command, and updating our HTML document to reference the JavaScript file under its new name:

```
[master *] $: mv js scripts
[master *] $: git status
On branch master
Changes to be committed:
  (use "git reset HEAD <file>..." to unstage)

    modified:   index.html
    new file:   js/site.js

Changes not staged for commit:
  (use "git add/rm <file>..." to update what will be
  committed)
  (use "git checkout -- <file>..." to discard
  changes in working directory)

    modified:   index.html
    deleted:    js/site.js

Untracked files:
  (use "git add <file>..." to include in what will
  be committed)

    scripts/
```

There's a lot going on here, and to Git's credit, this complicated status is relatively easy to follow and explain. First, we have the two changes we already staged—these remain staged, even though other changes to the working copy appear to have superseded them. If we committed right now, our scripts directory would be named js in the repository even though its name is scripts in our working copy.

In addition to the changes we've staged, Git also tells us about the newer updates that aren't yet staged. As before, because it no longer sees the js/site.js file under that name, it's reported deleted, and the scripts/site.js file that replaces it appears entirely new. We also see index.html listed twice, in two different states at once: modified and staged, but also modified and *unstaged*. git add saves a copy of the state of a file in Git's database for inclusion in a commit; here Git is trying to tell us that there's an even newer version of index.html than the one we previously staged.

But though this looks like a convoluted mess, `git add --all` resolves it neatly and quickly:

```
[master *] $: git add --all
[master *] $: git status
On branch master
Changes to be committed:
  (use "git reset HEAD <file>..." to unstage)

    modified:   index.html
    new file:   scripts/site.js
```

All of those seemingly contradictory changes have been condensed into just the two we want. Now we'll finally commit:

```
[master *] $: git commit -m "Add JavaScript"
[master 4af326c] Add JavaScript
 2 files changed, 1 insertion(+)
 create mode 100644 scripts/site.js
```

That's great, and much simpler, which begs the question: why not always use `git add --all`? Frankly, most of the time it's not only acceptable to use `--all`, but you're also almost always better off doing so. Not only will it save you time, but fewer commands means fewer opportunities to accidentally give Git a wrong signal, leading to confusion and heartbreak. With this option, you're telling Git to trust that the version of your project in the working copy is an accurate reflection of what you'd like to commit.

Even so, the fact that you *can*, but don't *have to*, commit everything that has changed in your local copy is one of Git's most powerful features. This is the power of the staging area: you can precisely control the scope of each commit, making each one broader or more focused, as your working style or the needs of your project demand.

3
BRANCHES

WE'VE TALKED A LOT SO FAR about "versions"—but what is a version, really? Webster defines it as "a particular form of something, differing in certain respects from an earlier form or other forms of the same type of thing." That is, versions can be *sequential* or *iterative*—representing the form of a thing as it changes over time—or they can differ in, well, some other way. The point is that a version isn't just a copy of a thing, but a copy that differs or has changed in some respect from some other copy.

In Git, sequential versioning—tracking the difference between a snapshot of your work and its earlier forms—is one of a commit's many jobs. Every commit includes a reference to its immediate predecessor, or *parent commit*; from that reference, Git can work backward and explain the entire chain of commits that came before it. In this sense, a commit represents both your work in a particular form, and its change into that form from a previous one.

That kind of version relationship is important for understanding where you've been, but not always helpful for understanding where you're going, or why. This is where *branches* come in handy. A branch is a virtual copy of your project—a project within your project—where you can make commits freely, in isolation from whatever else may be happening in your repository. Branches allow us to manage and work with *other* kinds of versions in Git—experiments, alternate takes, scratch pads—separately from the "official" copy of the work represented by the master branch.

Many people have valiantly tried to explain Git branches by comparing them to things, but no analogy does full justice to these beautiful, powerful buckets of pure information. The best way to understand branches is on their own terms, as a way of organizing and describing work. And the best way to explain *that* is for us to dive in and start growing some branches.

BRANCHING BASICS

Every Git repository starts out with a master branch, to which Git assigns the name master by default. Technically speaking, master is just a branch like any other; what makes it special is its conventional role as the primary, stable version of whatever project is stored in a repo. What "primary" or "stable" mean is largely up to you, and teams use their master branches in any number of ways. The only thing set in stone about master is that it's the first branch you'll work with. It won't be the last.

Before we create our first new branch, we'll view a list of all the branches on our local repository using the git branch command:

```
(master) $: git branch
* master
```

Here there's only one branch—master—and the asterisk tells us it's the current one.

Behind the scenes, a branch is little more than a human-friendly name that points to a particular commit. The diagram in **FIG 3.1** shows our current commit, 18ee782, outlined in red, with our current branch name (master) pointing to it. This particular branch is just a stack of commits, with 18ee782 at the

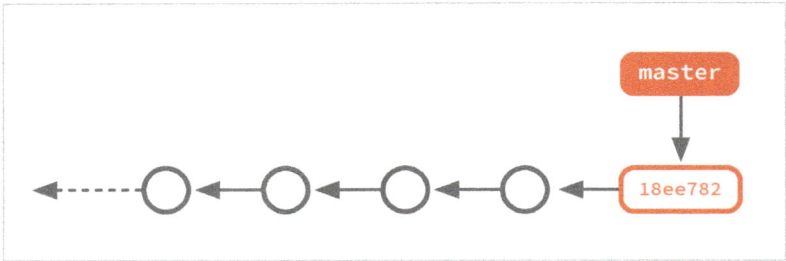

FIG 3.1: Our current commit is outlined in red, with the current branch name pointing to it.

top, or head, of the stack. (In keeping with the arboreal metaphor, a branch's head commit is sometimes also referred to as its "tip.") The stack is formed by following the head commit's chain of parent commits as far back as Git can go, all the way to the first commit in this repo.

When talking about branches, it's tempting to allude to them as places: when you add a commit, master is the "place" you're sending it to. As a working metaphor, this is fine most of the time. *Where did I commit the new footer links?* is a reasonable question that can often be answered with a branch name. There's no harm in treating branch names like folders in the metaphorical Trapper Keeper of your project.

It's not a perfect analogy, of course. While you can only put a sheet of paper in one file folder at a time, Git commits can belong to multiple branches—can be in multiple places—at the same time. But that's fine! Ultimately, a branch's most important role is as a signpost or bookmark, pointing you back to a particular version of your work, distinguishing master from, say, another branch named new-homepage. In this respect, branch names serve the same purpose as labels in Gmail. Just as an email can be labeled as both "Inbox" and "Notes from Mom" simultaneously, so too can a single commit be found in both the master and new-homepage branches. Branch names aren't so much destinations as they are labels, or signposts, that help you find certain commits.

STARTING A NEW BRANCH

For our next project, we've been asked to do something pretty significant: redesign our site's homepage. It may take some time and a lot of commits to get it right. But we don't want to publish our work before it's done, nor do we want to prevent any of our teammates from making changes to the site or the existing homepage while we're iterating.

We need a safe place to make potentially big changes without disrupting everything else going on.

The natural place for us to do this work is on a separate branch. In Git lingo, this is called a working branch or topic branch. Topic branches are distinguished from the master branch in that, well, they have a topic: the work that happens in them has a particular focus or goal, which is typically described by the branch name. Because we're making a new homepage, let's create a branch called new-homepage.

To create a new branch, just pass a branch name into the git branch command, like so:

```
(master) $: git branch new-homepage
```

This tells Git to create a new branch named new-homepage, using whatever commit you're currently on as a starting point. It doesn't matter which branch is set as the current one; git branch only cares which commit is at the head of that branch. Right now, commit 18ee782 is at the head of master, so this freshly minted new-homepage branch will also start out with 18ee782 at its head. But these two branches have no relationship to each other, aside from both having 18ee782 as a member.

Annoyingly, Git doesn't automatically switch you into the new branch when running git branch <branchname>. It creates a new branch, but leaves you with master set as the current branch, or (in Git terms) *checked out*. Left in this state, Git will have created your new branch, but your next commit will be to master.

Checking out a branch does two things. First, it resets your working copy to match whatever state is represented by the branch's head commit. Then, it sets the branch as current so that when you commit, any new commits you add will be added to that particular branch.

To switch branches, you'll use the git checkout command.

```
(master) $: git checkout new-homepage
Switched to branch 'new-homepage'
(new-homepage) $:
```

While you get used to working with branches, this two-command dance of creating and then switching into branches can quickly grow tiresome, so Git offers a handy shortcut: you can tell git checkout to create *and* switch to a new branch at the same time by passing in the -b option, like so:

```
(master) $: git checkout -b new-homepage
Switched to a new branch 'new-homepage'
(new-homepage) $:
```

Either way, once you've created and switched into a new branch, you should see it when you type git branch:

```
(new-homepage) $: git branch
  master
* new-homepage
```

The asterisk tells us that new-homepage has (correctly) been set as the current branch, meaning that our next commits will be sent here.

At the moment, because we haven't added any commits to either master or new-homepage, the two branches are literally identical—they represent two different names for the exact same commit.

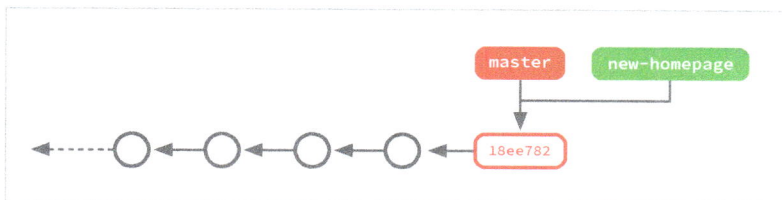

FIG 3.2: In the absence of new commits, these two branches are literally identical.

But part of the beauty in how Git handles branches is that, most of the time, you don't need to worry about that: these two things may be literally identical, but they are *logically* separate. Though master and new-homepage have identical contents, they are nonetheless two different logical copies of your work, not just two names for the same copy.

You do need to worry about knowing which branch is the current one, because even though these two branches are identical right now, they probably won't stay that way. By checking out the new-homepage branch, we're signaling our intention to diverge from the official timeline and to do some work that may or may not end up in the production version of our code.

Let's take a moment to savor some semantic details. As I said, the master and new-homepage branches are currently identical. In practice, that means that 18ee782 is the head commit for both branches. Therefore, it can be said that both branches *contain* that commit, along with its immediate parent and every commit that directly preceded it, all the way back to the initial commit in this repo. Since one of the branches we're talking about is master (and so far we haven't committed anywhere besides master), the full history of our project is contained within either or both of these branches.

I bring this up because we're about to make our first commit to a branch other than master. Having checked out the new-homepage branch, whatever we do next will be part of our topic branch, but not part of master.

Not *yet*.

OUT ON A LIMB

However grand our plans may be for this new, redesigned homepage, we have to start somewhere. For now, let's start with something easy: changing the background color on our website header from a blue gradient to flat gray, because flat design.

Let's make our CSS change and commit it:

```
(new-homepage *) $: git commit -am "Change
  background color on header"
[new-homepage b26b038] Change background color on
  header
 1 file changed, 1 insertion(+), 1 deletion(-)
```

Before we proceed, I'd like to call your attention to something. -am (as used in the preceding example) is actually a combination of two other options we've seen many times before: -a (to automatically add any changed files to this commit) and -m (specifying the commit message). Most command-line tools allow you to combine multiple options into a single one like this, prepended by a single dash. The only restriction is that just one of these options (m) can take an argument, and it needs to come last. This particular combination of options is very handy, as in many cases it allows you to commit some changes with only one command.

Having now added this commit, our master and working branches have diverged—the new-homepage branch has one commit that the master doesn't.

The diagram in **FIG 3.3** is more or less how Git sees the current state of our repo. The two branches perfectly overlap, except for that one new commit on the new-homepage branch.

That's how branches actually work: the new-homepage branch points to b26b038, master points to 18ee782, and b26b038 points to its parent commit, 18ee782.

Another way of looking at it is to see the two branches as discrete logical copies of the whole timeline that just happen to be mostly identical (**FIG 3.4**).

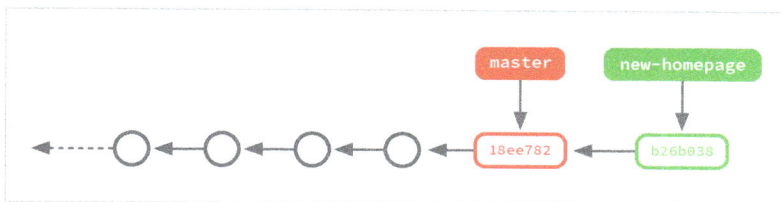

FIG 3.3: These two branches perfectly overlap except for one new commit.

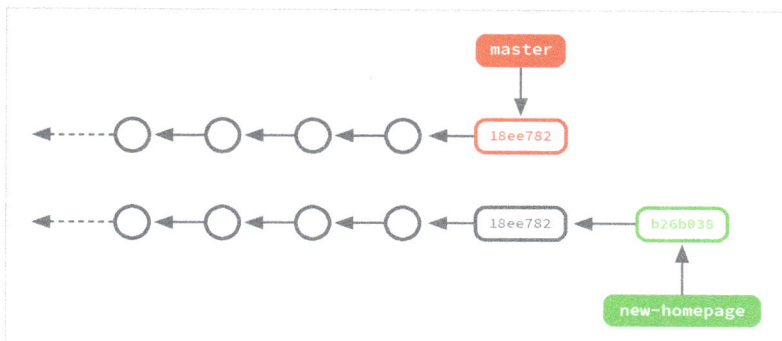

FIG 3.4: The two branches can be viewed as discrete logical copies that just happen to be mostly identical.

In technical terms, Git not only creates the commit but also moves the pointer for the current branch to the commit we just made. Put a little more simply: when you add a commit on a branch, that new commit supersedes the one that had been there before as the branch's head commit. Through the magic of parent commits, we can look backward and trace the lineage of a branch all the way back to the beginning, enabling us to produce graphs showing our two branches' shared histories. In practice, though, what matters about a given branch most of the time is which commit is on top of the stack.

It matters because the branch's latest head commit is the one your *next* commit will be based on. Working on Git branches is a lot like contributing to an *exquisite corpse* (http://bkaprt.com/gfh/03-01/), a kind of collaborative art project invented by the

Surrealists in the early 20th century, where each contributor sees only the last bits of whatever the previous person added. For example, one person might start a drawing, filling the first third of a sheet of paper. Then they fold the paper over, covering almost all of their work, before handing it off to someone else, who fills the next section of the paper, and so on. A modern version of exquisite corpse is Layer Tennis, where two designers pass a Photoshop file back and forth, adding a new layer with every turn (http://bkaprt.com/gfh/03-02/).

As you work and collaborate using Git, try not to worry too much about the whole system of commits, branches, and timelines. That stuff can be extremely valuable, and it's there for when you need it. But in the moment, feel free to focus on just getting from your last commit to your next one, step by incremental step.

NAMING BRANCHES

There truly are no hard and fast rules about what branches are for; as I mentioned, different teams tend to use branching in completely different ways.

There are some conventions, of course, but even those are fraught with ambiguity. For example, every Git repository has a master branch, and by convention master is meant to be the "prime" or "default" branch in your project. But it's up to you to decide what "prime" or "default" means in the context of your work.

Some projects, especially big open-source software projects like Ruby on Rails, use the master branch for all of the bleeding-edge work that'll go into their next release, periodically spinning off new "stable" branches to finalize and prepare code that'll end up in actual numbered versions of the Rails framework. Another approach, more common for websites and web-based applications, is for the master branch to represent the release version of the project that's deployed to your web servers, possibly many times a day.

What these two kinds of projects often have in common is how they use topic branches. There's a basic workflow around branching that, even though there is no single correct way to do things, seems to be how teams use Git to get work done most of the time.

First, someone checks out the latest version of master, and from that commit spins off a new branch named after whatever work they're planning to do, like in the new-homepage branch we created before. Our new-homepage branch is an example of a topic branch; exploring a new homepage design is the topic, and by branching off we're able to explore it freely without worrying about messing up the prime version in master.

Having branched off, we'll go off and work on our new idea or feature for a while, adding commits to the branch as we go. While all of this is happening, master can continue to evolve on its own, picking up new commits that aren't in your new branch. Nothing about the shared history of the new branch or master will change as a result of work on either branch, and all the new changes on one branch are isolated from changes on the other.

For instance, perhaps a new year is approaching, and you need to update the year in your site's copyright statement while you're in the middle of the homepage redesign. After you've committed your changes to the new-homepage branch, switching back to master is a git checkout away:

```
[new-homepage] $: git checkout master
Switched to branch 'master'

[master] $:
```

Once you've updated the year and committed that change, you can switch back just as easily with git checkout new-homepage.

The name you give to a branch should logically describe its reason for existing. Come up with a pithy label that identifies the work being done. A branch created to fix a problem with Chrome 32 might be called fix-chrome32-bug; it could also be named something more specific, like fix-chrome32-webfont-bug, or something more generic like bugfix. Choose a level of specificity that will distinguish this branch from others on your project without wasting space. Usable space in branch lists is at even more of a premium than in commit logs (although unlike with commits, it's possible to delete and clean out old branches that aren't being used). It's fine if branch names aren't completely descriptive—they just need to be descriptive *enough*.

MERGING

Sometimes, a branch will serve as a place to do work that you plan to throw away. That's one of the lovely things about branching in version control systems generally, and it's especially lovely in Git: branching is quick and cheap, and you're under no obligation to reconcile the version of your work in a branch with the one in master.

Most of the time, though, people use branches to work on things that they intend to fold back into the master copy eventually. Likewise, as master continues to evolve independently of whatever topic branches you're working on, you'll want to synchronize those changes (or at least some of them) with the ones in your branch. This is so that the version of your website saved in the branch is as fresh as possible—you wouldn't want to show off a version of your website where the homepage was new but everything else was obviously six months old—but also to ensure that your new work can merge easily back into master when it's ready.

Merging combines two (or more, but usually two) different branches of your project into a unified version that contains the unique attributes of both. On one side you have your master branch, containing the version of your site that's currently live on the web. Someone on your team discovers an error in the contact information, so they create a new topic branch named update-contact-info where the error is fixed. What you want now is a version of the master branch that includes the updated contact info from update-contact-info.

To do that, first check out the master branch on your local copy:

```
[update-contact-info] $: git checkout master
Switching to branch 'master'

[master] $:
```

Then use the git merge command to pull in changes from the other branch into your copy of master:

```
[master] $: git merge update-contact-info
Updating 286af1c..885e3ff
Fast-forward
```

Voilà! You now have a version of master that contains everything it had before, plus the amended contact info. Let's step back and examine exactly what just happened under the hood.

What merges are made of

In terms of how Git manages your data, when we say we want to end up with a version of our branch that includes all the stuff from another branch, we have two criteria that need to be met.

First, we want all of the commits we made on the other branch to be visible in our commit log (though their exact order can be flexible). Basically, we need for the history that led us to this moment to still have happened, and for the chain of ancestry to point back through both branches to include all of the work that was done so far.

Second, and more important, we need to end up with a copy of our project's files and folders that incorporates all of the changes from both branches. It may not surprise you to hear that the outcome of a merge is a commit and, as we've seen, every commit represents a complete snapshot of the whole project. The commit you end up with after a merge is no exception, which means someone or something needs to be responsible for constructing this new, unified version of the project directory.

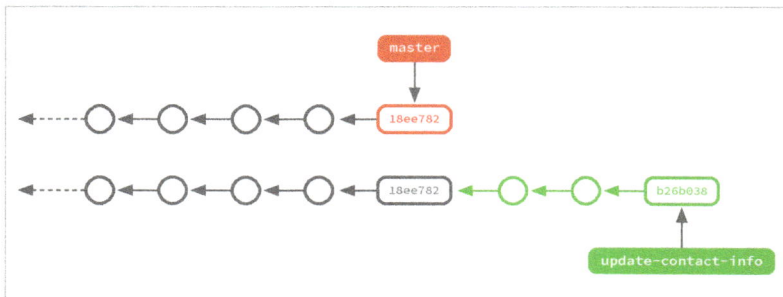

FIG 3.5: While the update-contact-info branch has a few more commits, master remains unchanged. Because only one side of the merger has changed, the merged state will be identical to what's in the update-contact-info branch.

Most of the time, the "something" responsible for constructing a merged copy of your project is Git, and it has a few strategies it uses to do that.

Fast-forwards

The simplest, easiest kind of merge is called a fast-forward, which is exactly what it sounds like. In our example of the updated contact page, nothing had changed in master since we branched off; everything that was different in the update-contact-info branch was stuff that was added after the last commit on our master branch (**FIG 3.5**).

Here, git merge doesn't have to do any work at all to figure out what the post-merge state of the project should look like, because only one side of the merger has changed. Therefore, the merged state will be identical to what's currently in the update-contact-info branch. All Git needs to do is move the master branch bookmark from its current commit to the head commit of the other branch (**FIG 3.6**).

Following a fast-forward, the two branches simply point to the same commit, making them once again identical in every way except their names.

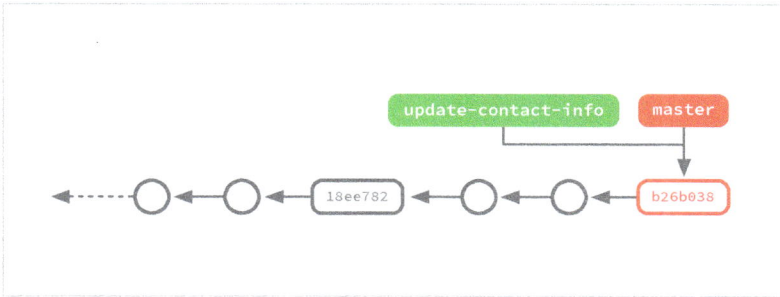

FIG 3.6: The master branch bookmark shifts from its current commit to the head commit of the other branch.

Merge commits

As smooth and elegant as fast-forwards are, they're not possible unless only one of the two branches has new commits, which—depending on the size of your team and how quickly things change in your project—may happen only rarely. The rest of the time, Git falls back to a "true merge," where it figures out what the combined state of the project should look like, creates a snapshot representing that new, merged version, and finally adds a special kind of commit—a *merge commit*—to tie everything together.

To return to our contact-info-updating example, let's say both master and update-contact-info have changed, each picking up one new commit since they branched off.

Because both branches have changes, Git has to do a little work to ensure the two branches can be merged safely. First, Git identifies changes between the head commits of each branch by looking for the first common ancestor of both versions, before working backward to understand what changed and in what order, using the common ancestor as a reference point. Then, Git compares each changed file in both branches against the reference point. When Git identifies a line that has changed in either branch, that line is carried forward for inclusion in the final, merged copy. As long as the branches don't both

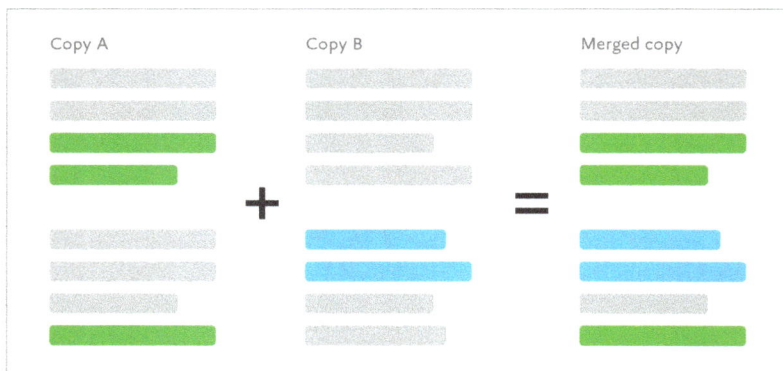

FIG 3.7: In a simple merge, Git finds only the changed lines from two branches and combines them to produce a third, merged snapshot.

contain changes to the same line, Git can still merge everything automatically.

Once the merged snapshot has been automatically generated, by default Git seals the deal and commits it for you, with an automatically generated commit message: "Merge branch 'update-contact-info' into master."

This is only a default, of course: you can pass the -no-commit option to git merge to ask Git just to generate and stage the merged-together version of your work, but wait for you to commit it yourself. One reason you might do this is to craft your own commit message, although the automatically generated one is almost always good enough. Another reason might be that you want to make some other changes in the same commit, or even to merge in several different branches in a single commit by running git merge --no-commit more than once. Doing so doesn't save you any work, though, and you lose the benefit of a merge commit marking when and where a merge occurred. All of this is to say: use the regular, automatic merge unless you have a reason not to.

Merge commits have a few unique properties. For example, unlike a normal commit, which has only one parent, merge

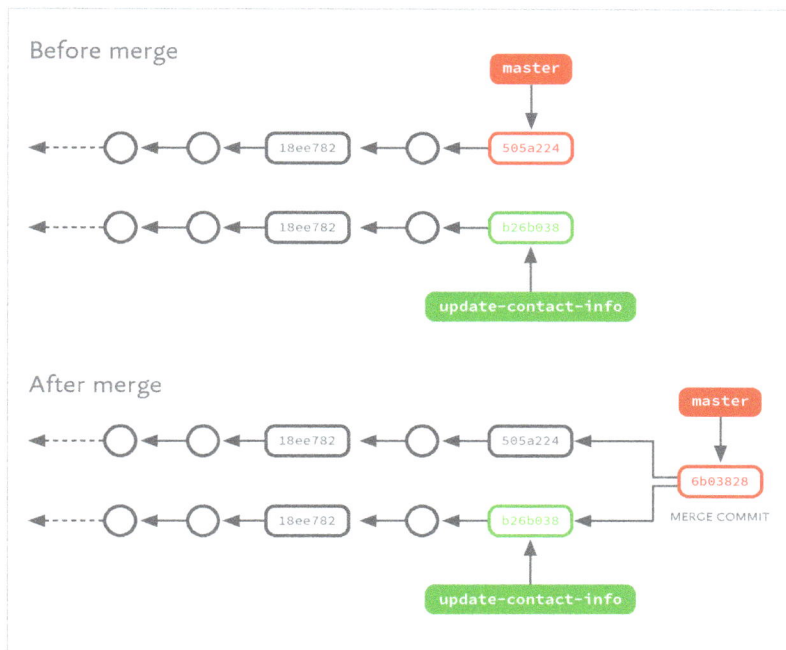

FIG 3.8: Because both master and update-contact-info have added a commit since their ancestor in common, 18ee782, Git creates a merge commit to tie them together, referencing both previous commits as parents.

commits can have two (*or more*, but usually just two) parents (**FIG 3.8**). In most respects, though, merge commits are like any other commit, and subject to the same rules.

For a few reasons, it's generally a bad idea for a topic branch to drift *too* far away from the latest version of master. Presuming here that topic branches are short-lived, and your ultimate goal is to merge them into master, keeping branches relatively up to date with master will make that eventual, final merge go more smoothly, and reduce the risk of dreaded merge conflicts.

So, from time to time, you'll want to merge new commits from master into your branch to bring it back up to date:

```
(new-homepage) $: git merge master
Updating c7038f8..1c4b16a
Fast-forward
 Makefile                                    |   7 ++
 Rakefile                                    |  15 ++--
```

Ideally, the final state of each working branch should differ from master only in ways that are relevant to its topic. For example, the new-homepage branch should have the changes needed to produce the homepage, but not any unrelated changes, and not any regressions to an older version of master. The same is true of update-contact-info, add-web-fonts, or any other topical branches you might work on.

HANDLING MERGE CONFLICTS

As we've seen, Git can often merge two branches together automatically. But sometimes it's not obvious how two branches should be merged together, in which case Git will ask for your help. This is one of the most annoying scenarios in Git, one that can seem really scary the first time you encounter it, but is actually not that bad: the dreaded *merge conflict*.

Most often, merge conflicts happen when two lines in a merge happen to overlap—that is, if two different versions are trying to change the same line of the same file. Under normal circumstances, Git will not try to resolve conflicts itself. Instead, it will do what it can, but after that it will stop and ask you to finish the merge commit yourself.

The process for resolving a merge conflict is very similar to the process for making a commit. To reinforce an earlier point, merge commits are ultimately just commits. They do have more than one parent, to reflect that they're combining two prior versions of your project into one unified whole. Otherwise, though, they follow the same rules as any other commit, including the process for creating them: first you need to stage changes you want included; then you commit.

To the extent that merge commits are different, it's in how much Git will try to do for you automatically before asking you to get involved. This, perversely, is one of the ways merge conflicts can throw off newcomers: by the time Git says it needs your help, the merge commit is often already mostly staged. It's as if someone else has started a commit and then left it for you to finish—and indeed, that's exactly what's happening.

Upon finding your branch in a conflicted state, if Git is able to successfully merge any files that have changed since the last commit on your current branch, those changes will be added to the staging area automatically. You generally don't need to worry about or do anything with these changes; they're good to go.

This is the part where the doctor says you'll feel a slight pinch: when Git can't automatically merge two copies of the same file, it will mark up your working copy and ask you to go through and manually choose which version of each change is the correct one. And when I say "manually choose," what I really mean is: Git will fill each conflicting file with arcane-looking gobbledygook called *conflict markers*, and your job is to go through each one and swap out the marked-up text for the version you want to end up with following the merge.

Let's dive into an example. One of our colleagues, Meghan, has recently been promoted from Director of Sales to Vice President, and we're working on a branch (promote-meghan) to add her new title to the company's About page. At the same time, someone else on our team is tweaking the HTML structure for the About page in their own branch (about-page-class-names), using Meghan's old title but changing all the markup around it. These two changes seem innocuous enough, but they're a recipe for a Git disaster.

Let's presume that about-page-class-names is merged into master first, so that the new HTML structure is now also part of master. Then, being conscientious Git users, we try to update our promote-meghan branch by running git merge master:

```
(promote-meghan) $: git merge master
Auto-merging about.html
CONFLICT (content): Merge conflict in about.html
Automatic merge failed; fix conflicts and then
  commit the result.
```

Oh, no! A merge conflict! Now, when we open up about. html, it looks like this:

```
<div class="team-member">
  <h2>Meghan Somebody</h2>
<<<<<<< HEAD
  <p class="title">Vice President, Market
  Development</p>
=======
  <p class="job-title">Director of Sales</p>
>>>>>>> master
</div>
```

The lines full of angle brackets are conflict markers, denoting the two versions of the conflicting line, which are separated here by the line full of equal signs. At the top of this conflict section, we have the version of the file in our current branch, identified here as HEAD. HEAD is Git's name for the "branch head pointer"; it's an alias for "the commit at the top of this particular branch," which in this case means the same thing as "the commit that is currently checked out in this working copy." This version reflects Meghan's new title, but shows the old CSS class name (class="title").

The bottom version is the one we're trying to merge in, identified here by its branch name (master). As you can see, the class name is up to date (class="job-title"), but the actual title is not. It's a subtle difference, but big enough for Git not to want to assume anything about which version is right, or how to correctly combine the two. Git doesn't know anything about job titles, and can't write HTML. It's relying on you, its human operator, to step in and craft the merged version.

To resolve this conflict, we need to replace all of this—everything between and including the angle-bracketed lines—with the version of the code that we want to wind up with following the merge:

```
<div class="team-member">
  <h2>Meghan Somebody</h2>
  <p class="job-title">Vice President, Market
  Development</p>
</div>
```

This handcrafted merge incorporates both the new class name *and* the new title. Next, we'll officially resolve the conflict by adding the new version to the staging area. Staging this change tells Git that we've officially signed off on this version being, well, the official one.

```
(promote-meghan *) $: git add about.html
```

We'll finish the merge by committing, with a message explaining what we did:

```
(promote-meghan *) $: git commit -m "Merge branch »
  master into promote-meghan, w/ resolved conflicts"
```

Although this commit also resolved a merge, *it's just a commit*. As mind-blowing as this may sound, you're under no obligation to use either of the two old versions of this line in this file. You could decide to resolve the merge by changing Meghan's title to give her an even bigger promotion to "CEO." (Don't be surprised to get a confused email from your actual CEO if you do that.) Whatever version of the file is staged when you make that last commit is the one that will "win" the merge.

As you can see, this mechanism for resolving conflicts is simple—perhaps a little *too* simple. If you're not thorough, it's possible to accidentally check in a file full of conflict markers, because Git will expect you to know that it's your job to find and remove them all. (If you *do* accidentally check in conflict markup, or anything else that shouldn't be in a file, *do not panic*. Just calmly fix whatever got messed up, stage your changes, and commit them. Git may be inscrutable, but at least it's consistent.)

Again, while merge conflicts are more annoying than truly scary, you're better off avoiding them whenever possible by merging in `master` regularly. This may seem ridiculous given that, in the last example, merging in `master` is precisely what tripped us up. But that particular merge conflict was inevitable given the nature of what had changed in the two branches. By merging in `master` now, we were able to deal with this one-line conflict when it was small and simple. Later on, we may have picked up other conflicts in addition to this one, and we'd have had to stop in our tracks while we painstakingly cleaned everything up by hand.

The goal of keeping master and topic branches up to date with each other isn't to prevent conflicts, but rather to make conflicts easier to manage by keeping the differences between two branches small. Most of the time, if you're judicious about merging, this kind of thing will never come up. Still, despite your best efforts, merge conflicts sometimes happen anyway. This scenario, where two different members of your team each make well-intentioned changes that happen to conflict with each other, is always possible. So it can be comforting to know that Git has a solution—and, as always, Git's solution is a commit.

4 REMOTES

SO FAR, ALL OF THE CHANGES we've made and committed live in one place: your computer. That's actually pretty neat: unlike some other version control systems that maintain their repository of committed versions exclusively on a server, requiring you to be online to commit changes, Git works offline by default. For solo projects, this means that you can benefit from powerful version control without having access to a server or needing to set up an account somewhere.

But working solo is not really why people come to Git. People usually come to Git because they want to collaborate.

There's an old saying (which, like the word "bug," is popularly attributed to Grace Hopper): *A ship in port is safe, but that's not what ships were built for.* Right now, all these commits you've been making are safe in the proverbial harbor that is your computer. Let's send them on a voyage.

A *remote repository*—as opposed to a local one on your computer—is a copy of a Git project that lives somewhere else: another computer on your network, someone else's computer somewhere else, an online service like GitHub—anywhere other than the directory you're looking at right now. In fact, strictly speaking, when I talk about your "local" repository, I'm referring only to the one you happen to be working with right now. You can even ask Git to push and pull changes to a second local copy stored in a different folder on your own computer, and that second copy would be considered a remote.

Remotes are one of Git's most successful abstractions. Unlike branches, which are wholly virtual copies of your project, each remote corresponds to an actual, physical copy of your repository with which you can exchange data. Most of the things you'll need to do to send and receive changes with a remote have been neatly wrapped up into two verbs: push and pull, which do more or less what you'd expect.

YOUR GIT HUB

Git's decentralized design allows you to push and pull changes between any two computers: if you wanted to, you could push commits from a branch on your computer directly to a branch on your teammate's computer, and vice versa. And while this seems cool, for most teams it introduces a lot of complexity without a lot of benefit.

Instead, many teams share code via Git through what I'll call the *hub model*. It's centralized in a good way: you and your team keep a shared copy of a project on a remote server (the hub), where it's accessible to everyone on the team. Each team member who joins the project copies (or *clones*) the project repository to their own computer, makes and commits changes there, and then uses the `git push` and `git pull` commands to synchronize their repo with the one stored on the server.

There's nothing special about remote repositories: they're just instances of the project, stored somewhere accessible so that you can push or pull commits to or from them. In theory, Git doesn't consider any one repository to be the canonical one for a given project, although in practice most teams have a single shared remote copy (often hosted on GitHub) that they consider the primary one—what Git conventionally calls the *origin*. As with master branches, what "primary" means is up to you, and the origin remote is what you make of it.

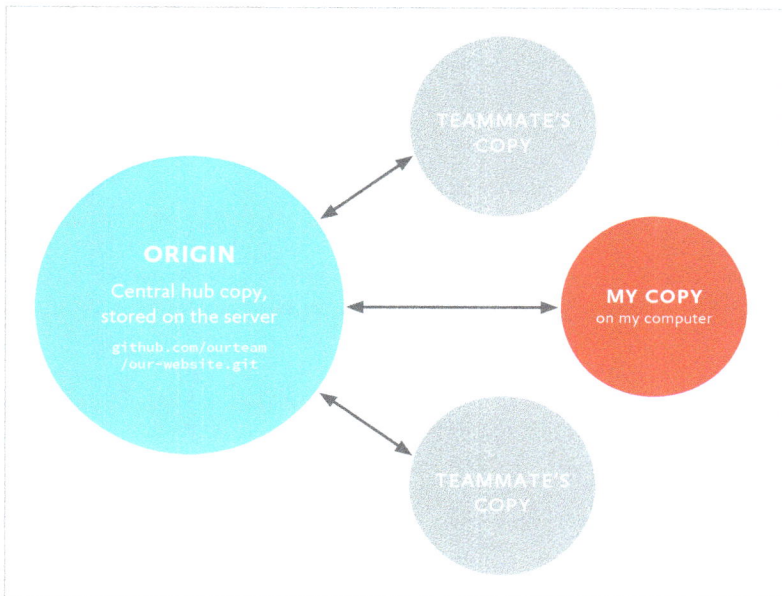

FIG 4.1: Following the hub model, members of your team synchronize their local copies with a shared, central copy, rather than with each other.

The hub model, though, views the origin remote as canonical, and so from the perspective of your team members, your changes aren't truly checked in until they're both committed *and* pushed to the server for others to access (**FIG 4.1**).

The hub also serves as a reliable backup of the code in the event that a contributor's own copy of the project gets corrupted or lost somehow, or if someone gets a new computer and needs to pull down a copy of his or her work. Rather than just copy files from one laptop to another, it's often easiest to re-clone the Git project from the hub to the new machine.

This of course presumes that the hub copy is never lost or corrupted, but Git's decentralized design helps us out here. Although the hub is the most canonical backup copy of your repo, *every* copy contains the complete history of your project. For work to be truly lost, it would have to disappear from *everyone's* computers, which is unlikely to say the least. In the improbable event that the hub becomes compromised, any local repo can be used to spawn a new remote.

WHAT LIVES ON THE SERVER?

Server-side repos are what are called "bare" repos, consisting only of the actual repository data (old versions, branches) and no working copy (which also means no staging area). A directory containing a bare Git repo is usually marked by appending `.git` to its name, as in our hypothetical `our-website.git`. The insides of a bare Git repo directory are virtually identical to what you'd find in the hidden `.git` directory in your local working directory, with subdirectories for objects, branch pointers, and other stuff Git needs.

Our server-side Git repo contains all of the commits that have been pushed to it, as well as its own set of branches. It's this additional, remote set of branches that can confuse the heck out of newcomers, because while it's natural for us to assume there's always a one-to-one relationship between a branch on our computer and one on the server, and while that's *usually* how it goes, Git doesn't require such a relationship and therefore doesn't enforce it. True to form, the main way Git compels you to deal with this loose coupling between local and server-side branches is by requiring you to be more specific in your commands.

For example, to push changes from one of your local branches to its twin on the server, it's often not enough to say just git push. Git may prefer that you say `git push <remotename><branchname>,` even if it seems to us like both the remote name and branch name can be inferred from context.

WHERE'S THE REMOTE?

A repository's location *relative to your local repository* is what qualifies it as a remote. In other words, a remote is ... elsewhere. Where is that, exactly?

For most of you, most of the time, your remote repository will live on GitHub. GitHub is the most popular hosting service for Git repositories by such a wide margin that it seems ridiculous to write this chapter as if there are alternatives. Even if your team never hosts projects on GitHub, you're certain to interact with a repo hosted on GitHub at some point in your work. To be sure, GitHub's service is both very inexpensive—free if your project is open source or at least browsable by the public, with cheap paid plans available if you need private code sharing—and very easy to use.

Many other options exist, however: both other hosted services and ways to self-host Git repositories. If you're not willing or able to manage your own servers, a hosted service like GitHub is the best choice—they do all the heavy lifting so you can focus on your project. But depending on the kind of work you're doing, or the kind of organization you're doing it for, you may have to ensure that your source code is stored in-house.

Fortunately, although different services may have different tools or interfaces for creating remote repositories, they all function the same way once they're set up.

ADDING YOUR FIRST REMOTE

You can pass a remote's URL as a parameter to each of the Git commands I just mentioned, which is fine if you only need to push or pull changes once and never again. Most of the time, though, you'll work with the same remotes over and over again during a project's lifespan. Instead of referring to remotes by their URLs, you can assign names to each remote you work with, and refer to it by its name instead.

At this point, of course, we have our own local copy of the project stored on our computer. But let's say we also have a remote Git repo (our-website.git) stored on our own server, gitforhumans.info, which we'd like to set up as the origin for our project. To do this, we'll use the git remote add command. Switch back to the Terminal and enter this command:

```
(master) $: git remote add origin »
   https://gitforhumans.info/our-website.git
```

I should point out that git remote is a new kind of command for us: one with *subcommands*. Whereas all the commands we've used so far have had just a single, one-word command name (e.g., git commit), all the commands related to configuring remotes are *namespaced*; that is, they're all two-word commands starting with remote: remote add, remote rm, and so on.

Typing just git remote, with no subcommand, instructs Git to show us a list of all the remotes we've added to this project, similar to how git branch shows a list of branches. As you can see, we only have one: origin:

```
$: git remote
origin
```

Note that if you started out by cloning the project to your computer from a remote server, using the git clone command, you'll find this step is already done for you. Repositories you clone from a remote always come preconfigured with that remote set as its origin.

Just as your project's primary branch has a conventional name (master), so does its primary remote: origin. (Notice how this simple yet effective naming convention reinforces the "hub" role for the remote repository: semantically, the *remote* is the origin for your project's code, and all of your local repos are just satellites orbiting the hub.)

Although origin is the conventional name, you can name remotes anything you want. Unless you have a really compelling reason, though, it's best to stick with convention and go with origin for your project's primary remote home.

Understanding remote URLs

Git supports three different networking protocols for moving commits and other data across networks: the Git protocol, SSH,

Git protocol	git://gitforhumans.info/hello.git
HTTPS	https://gitforhumans.info/hello.git
SSH	git@gitforhumans.info:hello.git

FIG 4.2: Each of these example URLs refers to the same repo—hello.git, on the server named gitforhumans.info—using each of Git's three protocol options. Most Git hosting services offer at least HTTPS and SSH.

and HTTP (**FIG 4.2**). In day-to-day practice, all three behave the same way. Git's protocols differ only in how you authenticate yourself with the server (that is, how you identify yourself and prove that you're you) and whether they support reading and writing changes, or just reading.

SSH (Secure Shell)

Git's SSH protocol is the exact same one many of us use to log in to remote servers every day. In fact, any SSH server you have access to can probably be used as a host for remote Git repositories. SSH remotes support both reading and writing, and you can use any authentication method SSH supports.

While Git doesn't have a default protocol, per se, SSH is so widely used for securely sharing Git repositories online that it has become a sort of default—a status Git reinforces by not requiring a protocol prefix for SSH URLs. Put another way, if you omit the protocol part of a URL, Git just assumes you mean it's SSH. GitHub's longtime default URL format for private repo access (e.g., git@github.com:username/reponame.git) uses SSH.

One drawback to the SSH protocol limits its usefulness in today's open-source ecosystem: it only works for private repositories, because SSH has no way of allowing someone to access resources without authentication. (It is a *secure* shell, after all.) Therefore, you may have to rely on a different protocol if you want to offer public access to some or all of your repos.

Fortunately, most Git hosts offer support for multiple proto-cols, so you can use HTTP to allow the public to download the latest stuff from your hot new JavaScript framework's master branch, while using SSH within your team to push commits to that branch.

SSH Git remotes, like many SSH servers, support logging in with a username and password, but it's more common to identify yourself using public key authentication, whereby you generate a unique, secure key pair and upload the public key to your account on a Git server such as GitHub, keeping the private key safe on your own computer. When you access a remote from that server, Git (or rather, SSH, working on Git's behalf) securely sends your private key, which acts as a kind of ID badge. (GitHub's help docs have a good summary of the process of creating key pairs: http://bkaprt.com/gfh/04-01/).

Git newcomers can find working with key pairs daunting and unfamiliar, but in return for this added complexity we get both security and (beyond the initial setup step) ease-of-use. Because each user generates a unique key pair on their own computer, it's easy for server administrators to manage pre-cisely who has access to which projects, especially when using hosting services like GitHub or Bitbucket, which offer great tools for managing users and keys.

HTTPS

This is, of course, the same HTTP we use to deliver content over the web. These days, many Git hosts (notably including GitHub) have made HTTPS URLs the default, partly because they're easier to use (you can authenticate HTTPS remotes with a username and password, rather than a SSH key), and partly because they're more versatile. Whereas SSH *must* be private, and *must* allow read and write access to your repositories, HTTPS offers more flexibility. You can allow anyone on the internet to pull down changes from your repo, while restricting push access to members of your own team.

Git protocol

Only the Git protocol is unique to Git, but these days it's rarely used, largely because it's read-only. This once made it a good choice for serving up public repos (say, on GitHub), and it paired nicely with SSH for projects that needed both public and private access. Today, however, HTTPS is a better choice.

Which should you use?

On purely private projects—if you're working on commercial software, say, rather than on open-source code—SSH is an excellent choice, and the most widely supported. That said, if you want the simplest, most consistent experience, I recommend using HTTPS whenever possible. Though SSH keys aren't hard to manage, they still aren't as easy to use as a username and password, and the fact that HTTPS URLs can be made public makes them easier to share.

You can learn more about Git's protocols and their pros and cons in Scott Chacon's excellent reference book *Pro Git*, which is available for free online (http://bkaprt.com/gfh/04-02/).

WORKING WITH REMOTE BRANCHES

This may sound obvious, but the main difference between working with branches and working with remotes is that remotes are on another computer. When you're working with branches, you're mainly concerned with managing different versions of your work stored on your own computer, within what I (and Git) call your *local copy*. With remotes, just as with branches, you're still managing different versions. In fact, your interactions with remotes will almost always be in the context of a branch. Once you've committed a change to a branch on your local repository, you can use git push to submit your copy of that branch—and all the new commits you've added—to the server. Whenever you need to refresh your copy of a branch with everyone else's latest changes, you use git pull.

Let's look at some examples of how you'll use these new commands in practice, starting with pushing.

Pushing changes

Having worked on our new homepage design for a while, we've discovered a bug in some JavaScript we've written. Someone else on the team has offered to help fix the problem, but first we need to get our changes into her copy of the project. To do this, we need to push the new-homepage branch from our computer to the server, where our teammate can find and pull from it.

The command we need here is git push <remote> <branch>. Again, Git wants us to be explicit here, listing exactly which remote we want to push to (origin), and which branch we want pushed (new-homepage). This is our first time accessing this particular remote, which is password-protected, so Git will prompt us to enter our credentials when we try to push or pull initially:

```
$: git push origin new-homepage
Username for 'https://gitforhumans.info': ddemaree
Password for 'https://ddemaree@gitforhumans.info':
Counting objects: 8, done.
Delta compression using up to 8 threads.
Compressing objects: 100% (6/6), done.
Writing objects: 100% (8/8), 743 bytes | 0 bytes/s,
  done.
Total 8 (delta 1), reused 0 (delta 0)
To https://gitforhumans.info/our-website.git
 * [new branch]      new-homepage -> new-homepage
```

Git does several things on our behalf when we push changes, and this long, convoluted response tells us about each one.

First, in the initial lines after the password prompt, Git packs up and sends our commits over the network:

```
Counting objects: 8, done.
Delta compression using up to 8 threads.
Compressing objects: 100% (6/6), done.
Writing objects: 100% (8/8), 743 bytes | 0 bytes/s,
   done.
Total 8 (delta 1), reused 0 (delta 0)
```

There's nothing *we* need to know in this block of text; it's saying that Git was able to pack up and send our data to the server successfully.

The next line is much more relevant for us:

```
* [new branch]      new-homepage -> new-homepage
```

Here, Git tells us that the remote server received our branch called new-homepage, and from it created a new branch on the server, also called new-homepage. Git doesn't require remote branches to have the same names as their local counterparts. However, for the sake of everyone's sanity, it's customary to keep branch names consistent.

Pulling changes

It's later in the day, and we've come back from getting a coffee to find that our teammate has submitted her changes, fixing the bugs in our JavaScript. Now it's time to get the changes she has committed to the new-homepage branch into our copy of the branch, by updating our branch using git pull <remote> <branch>.

Here again, Git asks us to be maddeningly explicit, specifying the remote and branch names:

```
$: git pull origin new-homepage
remote: Counting objects: 5, done.
remote: Compressing objects: 100% (3/3), done.
remote: Total 3 (delta 2), reused 0 (delta 0)
Unpacking objects: 100% (3/3), done.
From https://gitforhumans.info/our-website.git
 * branch          new-homepage    -> FETCH_HEAD
```

```
Updating fed3ac5..4f82376
Fast-forward
 carousel.js | 2 +-
 1 file changed, 1 insertion(+), 1 deletion(-)
```

As with git pull, the response includes several lines (beginning with remote:) that explain how data is being transferred between the two repos, which isn't very interesting. Let's skip past that, to where there *is* an interesting detail:

```
From https://gitforhumans.info/our-website.git
 * branch            new-homepage     -> FETCH_HEAD
```

Here, where you might expect Git to say it has pulled changes from the server's copy of new-homepage to our local copy of the same branch, the little ASCII arrow is instead pointing to something called FETCH_HEAD. To explain this, let me step back a bit and show you how pushes and pulls work behind the scenes.

Whenever you push or pull a branch, two things need to happen, both of which are reflected in this response from git pull.

First, Git needs to transfer a bunch of objects (that is, your commits and the files whose changes they're tracking) to or from the server. All those remote lines cover this part of the process, and the reason I can confidently tell you to ignore them is that it's exceedingly rare to run into problems there. The riskiest part of sending data between two computers is the possibility of one machine's data accidentally overwriting the other's without realizing it, resulting in data loss. One of the most wonderful aspects of Git's architecture is that it's virtually impossible for commits to conflict with each other, so sending or receiving objects is extremely safe. The worst side effect is that one copy ends up with too much data, but there's almost no risk of losing anything.

Once all the new commits are safe on your computer, we get to the second part: a merge:

```
Updating fed3ac5..4f82376
Fast-forward
 carousel.js | 2 +-
 1 file changed, 1 insertion(+), 1 deletion(-)
```

Because there weren't any other commits on our side since we handed this branch off to our colleague, Git is able to merge it back in as a simple fast-forward.

Git does this elaborate, three-step, copy-and-merging dance in order to ensure the safety of the work we've committed to our copy of new-homepage. Although copying a bunch of *commits* between computers is safe, as we've seen, merging *branches* sometimes creates conflicts that Git can't resolve on its own. What's more, even though with git pull we're asking Git to merge a server-side branch into one of our local ones, when pulling Git actually does all of the merging work on the local side, which means it needs to copy the server's new-homepage branch to somewhere on our computer before attempting to merge it into our branch. FETCH_HEAD is that somewhere. It's a temporary branch Git has created as a buffer, for purposes of merging in these newly fetched changes.

It's important to remember that merging is implicitly part of pulling (and, for that matter, pushing). Or, to flip it around, it's helpful to remember that both pushing and pulling are the remote form of merging. Both commands do the exact same job: they move a branch to another computer, then merge it into another branch.

Having pulled in changes from the server, our copy of new-homepage is now up to date, and we can get back to work.

Resolving merge conflicts: remote edition

As we've just seen, pulling in remote changes always ends in a merge. And, as we also know, sometimes merges result in conflicts. If anything, pushes and pulls are *more* conflict-prone than other kinds of merges, because there are frequently more people and changes involved over longer stretches of time.

And the risk of conflict is perhaps never greater than with the branch that, in most projects, changes most frequently: the origin's shared copy of master.

In the last chapter, I mentioned that it's a good idea to keep each branch you're working on that you eventually plan to merge into master updated with the latest changes in master. Put more simply, while working you should pull in the server's master branch regularly, to reduce the risk of merge conflicts, and to help keep any conflicts that do occur as minimal as possible. The command for this, if you haven't guessed, is git pull origin master, which works similarly no matter which branch you're in. Here we'll try to pull changes from origin/master into our own copy of new-homepage:

```
(new-homepage) $: git pull origin master
From https://gitforhumans.info/our-homepage.git
 * branch              master      -> FETCH_HEAD
Auto-merging about.html
CONFLICT (content): Merge conflict in about.html
Automatic merge failed; fix conflicts and then
  commit the result.
```

Oof! Once again, Git has been tripped up by a one-line difference on the About page. Just like when we changed Meghan's title, a commit on our branch changed some text in the heading (from "About our site" to "Our Team"), while a commit on master changed the surrounding markup. If we open up about.html, we'll see the conflicting change, surrounded by conflict notation:

```
<<<<<<< HEAD
<h1 class="big-heading">About our site</h1>
=======
<h1>Our Team</h1>
>>>>>>> 4f2d3c939deaf8f2824d2be04cb59b3f8342aedb
```

The good news is that the process for resolving a merge conflict is exactly the same whether it's the result of a local git merge, or an attempted git pull. Just like last time, we need to replace all of this with the version of the text we want to end up with in this branch:

```
<h1 class="big-heading">Our Team</h1>
```

Next, stage and commit the change to resolve the conflict in our local branch.

```
(new-homepage *) $: git add -A

(new-homepage *) $: git commit -m "Merge origin/ »
  master into new-homepage, with resolved conflicts"
```

Once this commit is done, our branch is fully up to date with the server's master. You can now push these changes, including the merge commit we just created, to the server's copy of this branch, or keep working.

While we're here, let me draw your attention to some new notation that I used in the commit message. Remote branch names often take the form remotename/branchname, as in origin/master (that is, the copy of master that lives on the origin remote) or testserver/bugfix (the bugfix branch on the testserver remote). Although remote branches almost always correspond to (or *track*) a branch on your own computer, they are technically separate branches, and this slash notation is a good way of distinguishing between the two copies without having to always say, as I did just now, "the copy of master on the origin remote."

Dealing with (push) rejection

While we've continued to work on the design for our new homepage, the teammate who helped us fix some JavaScript earlier has found and fixed another bug in our code. She committed and pushed her bug fix to the remote branch, but got pulled into a meeting before she could let us know she added some changes to our branch.

Meanwhile, we try to push some changes of our own to the branch and this happens:

```
(new-homepage) $: git push origin new-homepage
To https://gitforhumans.info/our-homepage.git
 ! [rejected]          new-homepage -> new-homepage
 (non-fast-forward)
error: failed to push some refs to ' https://
  gitforhumans.info/our-homepage.git'
hint: Updates were rejected because the tip of your
  current branch is behind
hint: its remote counterpart. Integrate the remote
  changes (e.g.
hint: 'git pull ...') before pushing again.
hint: See the 'Note about fast-forwards' in 'git
  push --help' for details.
```

Gulp. What causes Git to *reject* changes you're trying to push?

Generally speaking, server-side Git repos don't have working copies, staging areas, or, for that matter, human users who could help resolve merge conflicts. In fact, the lack of a working copy means remotes generally can't merge branches together *at all* if they require more than a simple fast-forward to merge in. The response from git push tells us as much:

```
 ! [rejected]          new-homepage -> new-homepage
 (non-fast-forward)
```

Fortunately, this situation is easily fixed by pulling changes down from the server, and then trying to push again. Fast-forwards work by moving a branch's HEAD pointer from the commit it's currently on to one of its direct descendants. When you pull in changes, the result is a merge commit—which happens to be a direct descendant of the remote branch's current head commit, and therefore qualifies for a fast-forward. *Boom.*

Long story short: if you want to avoid this kind of rejection, or any kind of Git shenanigans, always pull before you push to make sure your own local copy is up to date. There's rarely any harm to pulling changes, and frequently lots of benefit.

TRACKING BRANCHES

By default, nothing connects local and remote copies of a given branch. Even though they share the same name, and we know they logically represent the same piece of work, Git doesn't yet know that our local new-homepage and the server's new-homepage are in any way related, which is why we always have to tell git pull and git push which remote branches we want to work with. As elsewhere in Git, this need to be explicit can be annoying—but it's also powerful. You can potentially pull changes into new-homepage from *any* branch, on *any* remote. You could run git pull maniks-computer new-homepage-with-sass—where maniks-computer is your colleague Manik's laptop, and new-homepage-with-sass is a branch converting your CSS styles to Sass—and it would totally work.

Having said that, there is value in telling Git when the local and remote versions of a branch are related, by telling Git that a local branch is *tracking* its remote counterpart. For instance, when a branch is set up for tracking, you can push and pull changes by typing just git push or git pull, with no other arguments. Git will understand what you mean, and do the right thing.

The simplest way to set up a tracking relationship is to include the --set-upstream (or -u) option when invoking git push.

```
(new-homepage) $: git push -u origin new-homepage
Branch new-homepage set up to track remote branch
  new-homepage from origin.
Everything up-to-date
```

You only need to do this once per local branch, and if you forget to do it the first time you push, that's fine—you can do it any time, even if you have no new changes to push (indicated here by Git telling us everything is up to date).

MAKING FETCH HAPPEN

Git has one other remote-related command that's worth talking about. On the surface, git fetch sounds maddeningly similar to git pull. But whereas git pull works to pull down changes for just a single branch, git fetch can pull down *everything* from an entire remote repository at once.

You'll notice that when we run git fetch origin, the output is very familiar:

```
$: git fetch origin
remote: Counting objects: 5, done.
remote: Compressing objects: 100% (3/3), done.
remote: Total 3 (delta 2), reused 0 (delta 0)
Unpacking objects: 100% (3/3), done.
From https://gitforhumans.info/our-homepage.git
   9eb7cf6..fed3ac5  master       -> origin/master
```

First we see the same object-copying gobbledygook we've noticed several times already. However, at the bottom you can see that *something* has happened other than just copying a bunch of data from the server, something different from the merges or fast-forwards we've gotten used to. Specifically, Git has saved a copy of the server's master branch to a special, read-only branch on our local copy called origin/master.

Part of git fetch's job is to allow you to work offline. When I say git fetch works on whole repositories, I mean that: by default, it pulls down a snapshot of every branch in a remote so that you can compare, merge, or do any other sort of work with those branches without needing to be online the whole time. When Git was developed in 2005, before smartphones and airplane Wi-Fi were ubiquitous, if you wanted to work from a café or during a flight, you needed to have pulled down a copy of everything on to your computer. But you would not necessarily have wanted to take the extra step of merging every branch on the server with every branch on your computer. (For one thing, what if you had changes in a branch that weren't ready to merge in? What if some branches had conflicts?)

Git's solution is to keep track of the state of each branch in your remote repositories using a system of read-only, name-spaced branches on your local copy of the repo. I lied a little bit when I said earlier that `origin/*` was *just* a notation for identifying remote branches. `origin/master` is *also* an actual branch saved in your local copy of the repository. After fetching, you end up with copies of every single branch on the remote, even those that don't have a local equivalent on your copy of the project, such as branches started by other people.

For safety and speed, Git tries only to use the network for moving commits around, and does any *real* work on your computer. So, instead of trying to compare data on your computer with data on the server, Git instead makes a copy of what's on the server and lets you compare or merge against *that*. The `origin/master` branch represents the origin remote's master branch, pointing to whatever commit was at the head of that branch the last time you pulled it from the server.

Having these special offline copies of your remote branches can complicate matters rather than simplify them. For instance, we actually have *three* different branches called `master`: your local `master`, the remote's `master`, and your local `origin/master` that's supposed to—but isn't guaranteed to be—in sync with the remote `master`. Thankfully, branches like `origin/master` are read-only, and are designed to only ever represent a copy of what's on the server. Once you run `git fetch`, you can generally assume that each offline branch is an accurate representation of its twin on the server, and go from there.

CHECKING OUT AN EXISTING BRANCH

In most of the teams I've worked on, most branches have been owned by just one person, who was both the branch's original creator and usually also the one responsible for merging it into `master` when the work was complete. However, many projects are bigger than one person and take longer than a day to finish, and you may not be the first to be asked to work on a particular branch. You may even join a branch while someone else is still working on it, and many people may be contributing all at once. So how do you add a commit to someone else's branch?

First, you need to check it out. To do that, we'll use `git fetch` to pull down copies of all of the branches currently on the server:

```
[master] $: git fetch origin
remote: Counting objects: 5, done.
remote: Compressing objects: 100% (3/3), done.
remote: Total 3 (delta 2), reused 0 (delta 0)
Unpacking objects: 100% (3/3), done.
From https://gitforhumans.info/our-website.git
   9eb7cf6..fed3ac5  master        -> origin/master
   9eb7cf6..fed3ac5  new-homepage -> origin/
     new-homepage
```

Having fetched the latest stuff from origin, all our server branches are now available to us on our computer, even offline. We'll need to be online to push changes back up to the server, but we can do almost anything else until then.

For instance, we can ask Git to give us a list of every branch that existed on the server as of the last time we ran `git fetch`. Although by default the `git branch` command will only tell you what branches exist on your local copy, you can give it the `--remote` (`-r`) flag to ask it to instead show you all of the branches Git knows about from your remotes:

```
$: git branch --remote
origin/make-logo-bigger
origin/master
origin/new-homepage
```

Any of these can be checked out and worked on, or merged into one of your branches. `git pull origin master` is, in fact, just a shortcut for a `git fetch`, followed by `git merge origin/master`.

Next, we'll check out the branch we want to work on, helpfully called `make-logo-bigger`. We don't need to include the `origin/` prefix; if you're checking out a remote branch for the first time, Git will first check to see if you have a local branch by that name, and if not will automatically set up a new local branch to track the remote one.

```
[master] $: git checkout make-logo-bigger
Branch make-logo-bigger set up to track remote
  branch make-logo-bigger from origin.
Switched to a new branch 'make-logo-bigger'

[make-logo-bigger] $:
```

We've talked about where version control came from, and how to practice it on our own projects using Git. We now know how to make commits, create and merge branches, and synchronize our changes with other computers—and, by extension, with other people. Along the way, we've started to build up a history around our project.

Next, we'll dig into what we can do with all those commits now that we have them. Onward!

5 HISTORY

GIT IS AN EXCELLENT TOOL for synchronizing changes across all our computers, and that's how we almost always use it—to keep each other in sync with what we're doing *right now*. But although most of the time all we care about is the current version, or a few current ones, Git does a great job of storing and tracking *every* version of our project, and those other thousands of commits are still there, ripe for exploration.

Every commit you add to your repository contributes to the historical record of your project, so it's a good idea to make the best, most meaningful commits you can. In this final chapter, we'll look at some of Git's tools for inspecting your project's history, and how useful this history can be.

READING THE LOG

The simplest way to inspect your project's history is as an ordered list of commits. Git's primary tool for viewing such a list is the `git log` command. Hosting services like GitHub offer web-based tools for browsing your old commits; they do the same job as git log with a little more user-friendly panache. The advantage to learning git log is that, like the rest of Git's command-line interface, it works the same way no matter what computer or hosting service you use. And, unlike GitHub's commit search, it works offline.

By default, invoking git log will show a list of every commit in your project, from the current head commit all the way back to the beginning, in reverse chronological order, like so:

```
$: git log
commit 45b1ec87cd2fde95a110dfe3028e93d25c9af186
Author: David Demaree <david@demaree.me>
Date:   Fri Dec 26 16:28:41 2014 -0500

    Rename styles.css to main.css

commit bf8144d4690d3f6052dc7f42135e3e9944b96b5a
Author: David Demaree <david@demaree.me>
Date:   Thu Dec 25 13:24:25 2014 -0600

    Initial commit
```

The lines starting with commit denote, well, a commit, each of which takes up a few lines. The long string of letters and numbers are each commit's ID. Below that, we see the Author who made this commit (me), and the Date on which it was added. Finally, there's the commit's log message, shown indented underneath the metadata.

This is the history we've been crafting as we make changes and commits on the project. Logs like these are why commit messages exist, and why it's good for them to be short. Ideally, you should get a sense of how this project has evolved over time just from paging through git log's output and scanning the log messages.

The previous example shows the log's default output. But Git can tell you as much or as little about your commits as you want, in virtually any format. The --pretty option allows you to select from a number of predefined formats, or specify your own using a format string. Here's the built-in `oneline` format, which shows only the commit ID and log message on a single line:

```
$: git log --pretty=oneline
45b1ec87cd2fde95a110dfe3028e93d25c9af186 Rename »
    styles.css to main.css
bf8144d4690d3f6052dc7f42135e3e9944b96b5a Initial »
    commit
```

A complete list of available log formatters, and the syntax for defining your own as a format string, can be found in the Git documentation (http://bkaprt.com/gfh/05-01/). And Atlassian has published a thorough yet friendly tutorial showing all the options for formatting log output, including some brief explanations of why you would use certain formats (http://bkaprt.com/gfh/05-02/).

Specifying your starting point

As we learned earlier in the book, Git's concept of history is based on lineage: a commit contains a reference to one or more parent commits, which point to their parent commits, which point to *their* parent commits, all the way back to the beginning. `git log` appears to show a history of our project in reverse chronological order, but the chronology is kind of a side effect. What it's *really* doing is following the chain of parent commits to show you where your current commit comes from.

By default, the list it shows starts from the head commit on your current branch. If you have `master` checked out, it'll show you the complete ancestry of the top commit on `master`. You can, however, ask it to use any commit or branch as a starting point. For example, here we're asking it to show the history on our `new-homepage` branch:

```
$: git log new-homepage
```

Viewing a range of commits

You can even specify commit *ranges*; that is, you can ask for all log entries between two commit references so that you can see only what has changed between any two points in your project history.

This is best for seeing a list of what differs in a topic branch since we branched off. Here, we'll ask `git log` to show all the commits that have been added to `new-homepage` that aren't yet merged into `master`, using the `--oneline` option to make the log output easier to scan:

```
$: git log --oneline master..new-homepage
bce44eb Bigger navigation buttons
056c8fd Update hero area w/ new background image
7e53652 Make font loading async
```

Our range is listed here as `start..end`, or rather, `olderbranch..newerbranch`, or to be *really* pedantic, `branch..branchwithdifferentcommits`. You see, `git log` doesn't care about chronology and, as we know, there's nothing stopping `master` from having its own changes that aren't yet merged into a topic branch like `new-homepage`. The simplest way to explain what `git log branch-a..branch-b` does is that it shows you a list of all the commits in `branch-b` that aren't in `branch-a`. In the previous example, we see three commits from `new-homepage` that aren't yet merged into `master`.

What's really cool is that we can ask `git log` to show us a list the other way around—to give us a list of commits in `master` that aren't in `new-homepage`:

```
$: git log --oneline new-homepage..master
5514d53 Fix JavaScript bug on products page
4af326c Support for Microsoft Edge
```

This works with remote branches too, so you can find out if your local copy of a branch is trailing behind the server's copy. Here I'm asking git log to show me a list of commits on the server that I haven't pulled into my local branch yet, with a custom format string so I can see who made each commit:

```
$: git log --pretty='format:%h - %an: %s' »
  new-homepage..origin/new-homepage
635ce39 - Susan Oliver: Important legalese change
65ae00e - Stewart Colbert: Make many (JS) promises
```

If either side of the commit range is your current HEAD commit—that is, the commit that's currently checked out into your working copy—you can leave it blank. Here we've got new-homepage checked out, and we're asking to see a list of new commits from master:

```
[new-homepage] $: git log --oneline ..master
5514d53 Fix JavaScript bug on products page
4af326c Support for Microsoft Edge
```

This is exactly the same result as when we asked for a log on new-homepage..master earlier. Because new-homepage is checked out, Git infers that's the other side of the comparison we're asking for, saving us a little typing.

Filtering the log

Finally, as if that weren't enough, you can pass filtering options to git log to limit the list of commits, to show only a certain number of recent commits, or only those from within a certain date range, or only those added by a certain member of the team. For example, this command will show only commits in one of my repos that were added by me, that include the word "Heroku," that are more than three years old, and that changed the file called Gemfile:

```
$: git log --author=Demaree --grep=heroku »
  --oneline Gemfile
94d8ecb Gemfile tweaks to remove heroku
ccc5266 Merged heroku prep into master
```

This is just scratching the surface of what `git log` is capable of. To learn more about this powerful tool, Atlassian's `git log` tutorial offers a great summary of what's possible (http://bkaprt. com/gfh/05-03/).

THE LONG AND SHORT OF COMMIT IDS

The unique ID of a given commit is among the most important things you might use `git log` to look up. Git's commit IDs serve a few purposes, but the most important one is the most straight-forward: we use them to identify a commit, as in, "that change that messed up all the image tags happened in 65ae00e." So far, we've mostly seen commit IDs in a short form like that. Most of the time, for reasons I'll explain, the short form is fine, and the only form you'll need. Occasionally, though, you're likely to see commit IDs in their longer, unabridged form, like this:

```
65ae00edfe8a795199ed416a9d6df8c3cfe8bd0a
```

What's the difference? And why does Git use these weird-looking strings of letters and numbers to identify revi-sions, instead of just a number?

As covered in the last chapter, even though many of us use Git in a centralized way, Git is designed to be *decentralized*. Every one of our computers has its own copy of the repository, which can evolve independently from the others. You and I can each make changes and commit them to a branch while offline, and neither of us needs to know what the other is doing until later, when we sync our local copies with a remote. As we make those commits, Git needs to be able to assign an identifying name or number to each one, but Git can't know ahead of time whether some other computer has already used that name or number.

What's more, Git's design values stability and data integrity above all else. In a 2007 presentation, Linus Torvalds talked about the need for version control systems to look after the veracity of the data under their care, and talked up Git's features for ensuring correct data (http://bkaprt.com/gfh/05-04/):

If you have disk corruption, if you have any kind of problems at all, Git will notice them ... I guarantee you, if you put your data in Git, you can trust the fact that five years later, after it was converted from your hard disk to DVD to whatever new technology and you copied it along, five years later you can verify that the data you get back out is the exact same data you put in.

Git solves both problems by creating and using IDs based on the *contents* of each commit, rather than arbitrarily assigning each one a name or number. Technically, commit IDs aren't identifiers so much as *checksums*, a kind of digital fingerprint, typically used to validate data that has been transmitted over a network. You'll often see a list of checksums alongside software builds, so people downloading, say, a prerelease build of Windows can verify that the downloaded file is complete, and hasn't been tampered with.

When you make a commit, Git takes everything that constitutes the body of the commit—your name and email address, the current date and time, the commit message, references to any parent commits and the current project snapshot—and runs them through the hashing function to generate that 40-character string. The result is a value that's virtually guaranteed to uniquely identify a given commit. That's true even if the same commit is made on two different computers. Two identical commits will have identical hashes, and therefore identical IDs, regardless of which computer added them to the repo. Conversely, commits that differ in any way—even just by having a different author—are guaranteed to have different IDs; therefore, each hash is guaranteed to uniquely identify a single commit.

While these long hashes help smooth collaboration, by making it easier to swap commits between computers, they also create a new problem for us. Because they are so long, reading and writing them can be unwieldy. Fortunately, even if you provide only a fragment of the full commit ID, Git is smart enough to figure out what commit you want, as long as the short ID is at least four characters long, and unique within your repo.

For instance, the commit ID I showed at the start of this section could be shortened to as few as four characters (65ae) without overlapping with any other commits in that project. In fact, in most Git repositories, a seven-character ID like 65ae00e is sufficient to uniquely identify any commit, even in repositories with tens of thousands of commits. For that reason, Git will frequently use short IDs in its responses to you rather than the longer form.

In the rare scenarios when two short IDs overlap, Git is also smart enough to handle things gracefully by automatically adding digits to the short IDs it prints out. In the Linux kernel project (http://bkaprt.com/gfh/01-02/), for instance—perhaps the oldest Git repository and certainly one of the biggest—it turns out that seven characters are not enough to avoid overlapping IDs, but eleven digits *do* work, so Git automatically switches its short ID format to use the fewest digits that will still be unique across the whole project.

COMMIT MESSAGES

Git and tools like GitHub offer many ways (some of which we'll look at later in this chapter) to view what actually changed in a commit. But a well-crafted commit message can save you from having to use those tools by neatly (and succinctly) summarizing what changed.

The log message is arguably the most important part of a commit, because it's the only place that captures not only what was changed, but *why*.

What goes into a good message? First, it needs to be short, and not just because brevity is the soul of wit. Most of the time, you'll be viewing commit messages in the context of Git's commit log, where there's often not a lot of space to display text.

Think of the commit log as a newsfeed for your project, in which the log message is the headline for each commit. Have you ever skimmed the headlines in a newspaper (or, for a more current example, BuzzFeed) and come away thinking you'd gotten a summary of what was happening in the world? A good headline doesn't have to tell the whole story, but it should tell you enough to know what the story is about before you read it.

If you're working by yourself, or closely with one or two collaborators, the log may seem interesting just for historical purposes, because you would have been there for most of the commits. But in Git repositories with a lot of collaborators, the commit log can be more valuable as a way of knowing what happened when you weren't looking.

Commit messages *can*, strictly speaking, span multiple lines, and can be as long or as detailed as you want. Git doesn't place any hard limit on what goes into a commit message, and in fact, if a given commit does call for additional context, you can add additional paragraphs to a message, like so:

```
Updated Ruby on Rails version because security

Bumped Rails version to 3.2.11 to fix JSON »
  security bug.
See also http://weblog.rubyonrails.org/2013/1/8/ »
  Rails-3-2-11-3-1-10-3-0-19-and-2-3-15-have-been- »
  released/
```

Note that although this message contains a lot more context than just one line, the first line is important because only the first line will be shown in the log:

```
commit f0c8f185e677026f0832a9c13ab72322773ad9cf
Author: David Demaree <david@demaree.me>
Date:   Sat Jan 3 15:49:03 2013 -0500

    Updated Ruby on Rails version because security
```

Like a good headline, the first line here summarizes the reason for the commit; the rest of the message goes into more detail.

Writing commit messages in your favorite text editor

Although the examples in this book all have you type your message inline, using the `--message` or `-m` argument to `git commit`, you may be more comfortable writing in your preferred text editor. Git integrates nicely with many popular editors, both on the command line (e.g., Vim, Emacs) or more modern, graphical apps like Atom, Sublime Text, or TextMate. With an editor configured, you can omit the `--message` flag and Git will hand off a draft commit message to that other program for authoring. When you're done, you can usually just close the window and Git will automatically pick up the message you entered.

To take advantage of this sweet integration, first you'll need to configure Git to use your editor (specifically, your editor's command-line program, if it has one). Here, I'm telling Git to hand off commit messages to Atom:

```
$: git config --global core.editor "atom --wait"
```

Every text editor has a slightly different set of arguments or options to pass in to integrate nicely with Git. (As you can see here, we had to pass the `--wait` option to Atom to get it to work.) GitHub's help documentation has a good, brief guide to setting up several popular editors (http://bkaprt.com/gfh/05-05/).

Elements of commit message style

There are few hard rules for crafting effective commit messages—just lots of guidelines and good practices, which, if you were to try to follow all of them all of the time, would quickly tie your mind in knots.

To ease the way, here are a few guidelines I'd recommend always following.

Be useful

The purpose of a commit message is to summarize a change. But the purpose of summarizing a change is to help you and your team understand what is going on in your project. The information you put into a message, therefore, should be valuable and useful to the people who will read it.

As fun as it is to use the commit message space for cursing—at a bug, or Git, or your own clumsiness—avoid editorializing. Avoid the temptation to write a commit message like "Aaaaahhh stupid bugs." Instead, take a deep breath, grab a coffee or some herbal tea or do whatever you need to do to clear your head. Then write a message that describes *what changed in the commit*, as clearly and succinctly as you can.

In addition to a short, clear description, when a commit is relevant to some piece of information in another system—for instance, if it fixes a bug logged in your bug tracker—it's also common to include the issue or bug number, like so:

```
Replace jQuery onReady listener with plain JS; »
  fixes #1357
```

Some bug trackers (including the one built into every GitHub project) can even be hooked into Git so that commit messages like this one will automatically mark the bug numbered `1357` as done as soon as the commit with this message is merged into `master`.

Be detailed (enough)

As a recovering software engineer, I understand the temptation to fill the commit message—and emails, and status reports, and stand-up meetings—with nerdy details. I *love* nerdy details. However, while some details are important for understanding a change, there's almost always a more general reason for a change that can be explained more succinctly. Besides, there's often not enough room to list every single detail about a change and still yield a commit log that's easy to scan in a Terminal window. Finding simpler ways to describe something doesn't just make the changes you've made more comprehensible to your teammates; it's also a great way to save space.

A good rule of thumb is to keep the "subject" portion of your commit messages to one line, or about 70 characters. If there are important details worth including in the message, but that don't need to be in the subject line, remember you can still include them as a separate paragraph.

Be consistent

However you and your colleagues decide to write commit messages, your commit log will be more valuable if you all try to follow a similar set of rules. Commit messages are too short to require an elaborate style guide, but having a conversation to establish some conventions, or making a short wiki page with some examples of particularly good (or bad) commit messages, will help things run more smoothly.

Use the active voice

The commit log isn't a list of static things; it's a list of *changes*. It's a list of *actions* you (or someone) have taken that have resulted in versions of your work. Although it may be tempting to use a commit message to label a version of the work—"Version 1.0," "Jan 24th deliverable"—there are other, better ways of doing that. Besides, it's all too easy to end up in an embarrassing situation like this:

```
# Making the last homepage update before releasing
  the new site
$: git commit -m "Version 1.0"

# Ten minutes later, after discovering a typo in
  your CSS
$: git commit -m "Version 1.0 (really)"

# Forty minutes later, after discovering another
  typo
$: git commit -m "Version 1.0 (oh FFS)"
```

Describing changes is not only the most correct format for a commit message, but it's also one of the easiest rules to stick to. Rather than concern yourself with abstract questions like whether a given commit is the release version of a thing, you can focus on a much simpler story: *I just did a thing, and this is the thing I just did.*

Those "Version 1.0" commits, therefore, could be described much more simply and accurately:

```
$: git commit -m "Update homepage for launch"
$: git commit -m "Fix typo in screen.scss"
$: git commit -m "Fix misspelled name on about page"
```

I also recommend picking a tense and sticking with it, for consistency's sake. I tend to use the imperative present tense to describe commits: *Fix misspelled name on About page* rather than *fixed* or *fixing*. There's nothing wrong with *fixed* or *fixing*, except that they're slightly longer. If another style works better for you or your team, go for it—just try to go for it consistently.

What happens if your commit message style isn't consistent? Your Git repo will collapse into itself and all of your work will be ruined. *Kidding!* People are fallible, lapses will happen, and a little bit of nonsense in your logs is inevitable. Note, though, that following style rules like these gets easier the more practice you get. Aim to write the best commit messages you can, and your logs will be better and more valuable for it.

For more on the art of writing commit messages, check out Tim Pope's "A Note About Git Commit Messages" (http://bkaprt.com/gfh/05-06/) and Chris Beams's "How To Write A Git Commit Message" (http://bkaprt.com/gfh/05-07/).

MAKING GOOD COMMITS

On 24 Ways, Emma Jane Westby wrote that "commits should really contain whole ideas of completed work" (http://bkaprt.com/gfh/05-08/). For us humans, the job of a commit is to bundle changes into logical chunks. Sometimes, the logic behind a particular set of changes is as simple as: "This is when I, the developer, felt it made sense to save my progress". But sometimes there's more of a story—more meaning—behind a change.

For a software tool so concerned with keeping your data clean and consistent, Git is remarkably flexible about exactly what you commit and when. One really cool (and potentially confusing) thing about Git is that it doesn't require you to stage or commit everything you've changed all at once. Git lets you move *some* changed files—or even changed *parts* of files—down the path from working copy to committed, while leaving other stuff unstaged or uncommitted. If you make three sort-of-unrelated changes to a single stylesheet file, you can commit each of the changes separately, or together, as you see fit.

Let's say you're working on a project for which you've changed both a JavaScript file and your project README, for unrelated reasons. Here's our status:

```
[master] $: git status
# On branch master
#
# Initial commit
#
# Untracked files:
#   (use "git add <file>..." to include in what will
  be committed)
#
#       README.md
#       site.js
```

The simplest thing would be to commit both files at the same time, with a joint log message like "Add onReady event listener; update README." But if committing the two changes separately is more meaningful, and provides more context for your logs, Git makes it relatively easy to do that.

First, let's stage and commit one of our two changes:

```
$ git add site.js
$ git commit -m "Add onReady event listener"
[master 591672e] Add onReady event listener
 1 file changed, 3 insertions(+)
```

After we do that, our README is still modified and unstaged, ready for us to commit separately:

```
$ git add README.md
$ git commit -m "Update README"
[master 96406dd] Update README
 1 file changed, 1 insertion(+)
```

Now, if we check the log, we'll be at least a little more confident that each entry in it—each commit—represents a single, complete idea.

This is a hard thing to do perfectly all the time and, like a lot of other best practices, commits that are perfect, single units of work, wrapped up in a perfectly worded commit message, are the exception rather than the rule. Don't beat yourself up for a big, messy commit with a vague label like "fixed the header"— just know that better is possible and aim for it when you can.

COMPARING COMMITS

We've talked a lot about versions, states, and how changes add up incrementally over time. When we deal with our work one commit at a time, we're encouraged to think beyond the state of our work *right now* and consider the state it was in yesterday, and the state it will be in tomorrow. From writing commit messages and deciding what should go into a commit, we're prompted to think about how we describe the actions we make as we make them, which eventually lends itself to a more thoughtful, considered approach to work.

Most of all, Git asks us to treat changes to our projects—more formally, the transitions between states represented by commits—as actual events that occurred. Each commit represents not only a snapshot of our whole project, but (except for the first one, of course) *also* a change from a previous commit. Eventually, once you start thinking and working in versions, you will want or need to compare the versions to see, specifically, what has changed. A commit message can give you a summary, but

Git also offers a handy way to actually inspect the differences between two commits.

`git diff` (short for "difference") shows the changes between two versions of your project, or two versions of a given file or files. In this way it's a lot like `git log`, and in fact you can choose to see diff information in your log output if you want. In addition to comparing committed versions, if you've made uncommitted changes, you can use `git diff` to show you everything that's different between your working tree and the last commit.

Here, `git diff` shows us a simple change to the README file we were looking at before:

```
$: git diff
diff --git a/README.md b/README.md
index 0c0a11f..48fb805 100644
--- a/README.md
+++ b/README.md
@@ -1 +1,3 @@
-# My Project
\ No newline at end of file
+# My Project
+
+This is a project managed by Git.
\ No newline at end of file
```

Admittedly, this is not the easiest thing to read. `git diff`'s default output is generated using a Unix comparison tool (itself called "diff") originally developed in the early 1970s, and displayed using a paging program called "less", whose job it is to display texts longer than your Terminal window. (I should point out that this is a rather simple example. Most of the time, your diffs will be longer and more complex.)

Here's what's going on in *this* diff: the lines starting with dashes (-) are ones that we've deleted since the last commit; the lines starting with plus signs (+) have been added.

Diffs (like Git generally) focus on changed *lines* in your files, and changing even one character in a line will cause Git to consider the line changed. Also, just as renaming a file is seen by

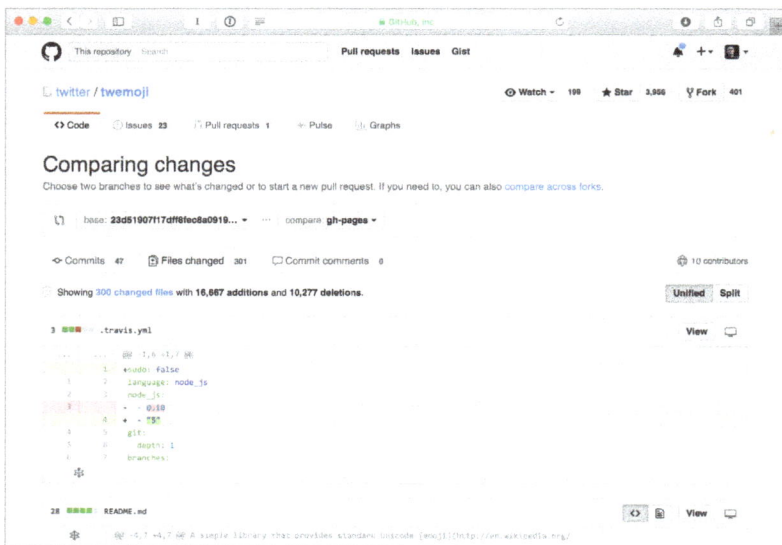

FIG 5.1: For projects hosted on GitHub, compare views are a great way to view—and share—colorful, easy-to-read diffs between two commits.

Git as a combination of deleting the old file and adding a new one, changing a line is seen as both a deletion and an addition. You can see that in this diff: the main headline is present in *both* versions, as a deletion *and* an addition. What changed? In the committed version, the headline wasn't followed by a line break. (Yes, even adding a line break is enough for the line to be marked as changed.)

Diffs can be incredibly useful, but unless you're very comfortable with the Unix diff format, they're also one of the few things in Git for which I wholeheartedly endorse using a GUI tool, either as an app on your computer or as part of a hosting service like GitHub. For projects hosted on GitHub or GitHub Enterprise, every repo has a compare view (accessible by appending /compare to your project's URL) that does a

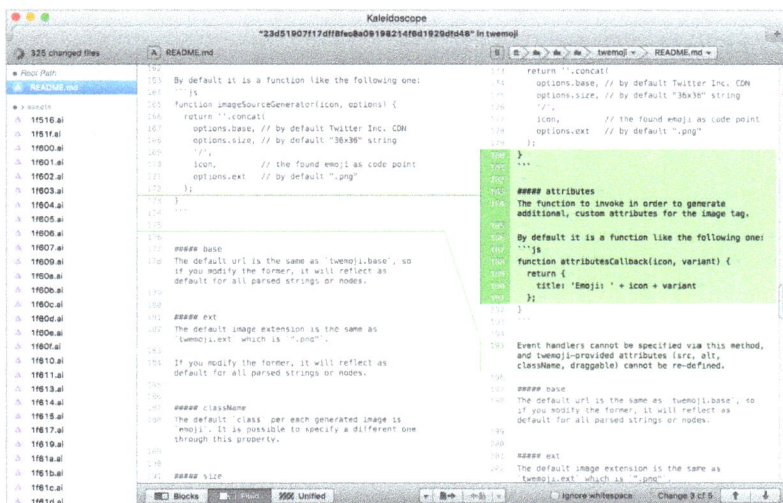

FIG 5.2: Git can integrate with visual file-comparison apps, like Black Pixel's excellent Kaleidoscope, that display diffs in a colorful, easy to navigate format.

great job of summarizing changes in an easy-to-use visual format (**FIG 5.1**).

Mac users might also consider Black Pixel's app Kaleidoscope (http://bkaprt.com/gfh/05-09/). Kaleidoscope is a general-purpose file comparison tool that can be used to compare any two files, regardless of whether they're managed by Git. That said, it offers great integration with Git, including an easy-to-use setup tool that configures Git to open Kaleidoscope for diffs via the git difftool command.

Git does offer a simpler diff format that is quite easy for humans to read, if we return to the command line: the "diff stat," which reduces a whole diff to a list of the files that differ between two versions, marked up to indicate how they've changed.

Here, I ask git diff to show stats for the difference between the current HEAD commit on Typekit's Web Font Loader repo and the one before it (HEAD~1):

```
$ git diff --stat HEAD~1
CHANGELOG               | 3 +++
lib/webfontloader.rb    | 2 +-
webfontloader.gemspec   | 2 +-
webfontloader.js        | 4 ++--
4 files changed, 7 insertions(+), 4 deletions(-)
```

While more concise than the full diff, the diff stats offer a good summary of the changes in this commit. Each line shows a file that was changed in this commit, like `lib/webfontloader.rb`. Next to it, separated by a pipe (|) character, are the stats for that file: two changes (one addition and one deletion). Knowing Git as well as we do now, observing that this is just one commit's worth of changes, we can infer that it may have been a one-line edit, such as a change in version number.

From here, if we need more information, we can request a full diff of a particular file (using `git diff HEAD~1 webfontloader.js`), or a set of files (by passing in multiple file names), or the whole project. We can also ask for stats covering commits made over a much broader span of time:

```
$: git diff -stat HEAD~15
.travis.yml                            |   5 +-
CHANGELOG                              |  12 ++
README.md                              |  24 +--
lib/webfontloader.rb                   |   2 +-
package.json                           |   5 +-
spec/core/fontwatcher_spec.js          |   3 -
spec/core/fontwatchrunner_spec.js      | 441 ++++++++++--
     ------------
src/core/domhelper.js                  |  26 ++-
src/core/fontruler.js                  |   2 +-
src/core/fontwatcher.js                |  22 +-
src/core/fontwatchrunner.js            |  33 +--
webfontloader.gemspec                  |   6 +-
webfontloader.js                       |  42 ++--
13 files changed, 207 insertions(+), 416
 deletions(-)
```

While this seems to roll up fifty commits' worth of changes into a single summary, this is a good place to clarify that `git diff` (including the stats view) only compares two commits at a time. `HEAD~50` doesn't represent the last fifty commits, just the one commit that's fifty steps back in your chain of ancestry. But let's also remember that every commit is a full snapshot of your project, and that every commit builds on the one before it. Logically, seeing the differences between your current commit and its fiftieth parent should be roughly the same—certainly the same in spirit—as seeing a summary of your last fifty changes, because those changes should all still be around in your current commit.

If you find the stats valuable, you can even include them in your git log output using the `--stat` option. Here, I'm asking git log to show me a log that includes stats, *plus* a custom format for the log entries, limited to the changes since the commit before last.

```
$ git log --stat --pretty=format:"%h (%an) %s" »
  HEAD~1..
d08a7f2 (Bram Stein) Release 1.5.10
 CHANGELOG              | 3 +++
 lib/webfontloader.rb   | 2 +-
 webfontloader.gemspec  | 2 +-
 webfontloader.js       | 4 ++--
 4 files changed, 7 insertions(+), 4 deletions(-)
```

TAGGING COMMITS

In addition to all the other kinds of references we've seen—long and short commit IDs, branch names, and the `HEAD` pointer—commits can be given permanent, human-friendly names, called *tags*. Tags are a lot like branches in that they assign human-readable names to a particular commit. But unlike branches, whose names, though consistent, float as the `HEAD` commit on each branch change, tags *always* reference a specific commit, to mark moments in history that are interesting or significant.

Depending on your project, you may never use tags, or you may use them a lot. Unlike branches, which are central to almost every Git workflow I've seen, tags have no intrinsic meaning or intended use, so many projects never use them. For web sites and applications, tags' value may depend entirely on how you release code to your production servers. Many teams deploy by just updating the servers with the latest stuff from `master`; they control what code goes out to the public by laying down rules about when and how commits can be merged in, and quality checks to ensure everything in `master` is always production-ready. For most of us, branches are not just simpler but more meaningful—the branch name `master` doesn't just reference a commit, it references the *latest* commit on a certain line of work. Branch names change less often, and so involve less work.

Git tags are commonly used for software libraries or frameworks that are shipped in numbered versions. For instance, the code for version 4.2.0 of the Ruby on Rails framework matches up with the `rel-4.2.0` tag on their Git repo (http://bkaprt.com/gfh/05-10/), which in turn points to commit `7847a19`, whose message, helpfully, is "Preparing for 4.2.0 release." The official 4.2.0 release is in the form of a Ruby package hosted on rubygems.org; the tag serves to connect that package with the commit used to produce it.

To tag a commit, you'll use the aptly named `git tag` command. It always takes as its first parameter the tag name, which can be any string. Here, we'll tag the current commit on our current branch with the name `fhqwhgads`. (If this seems like a bizarre example, you should know I once worked on a team that tagged our biweekly website releases after our favorite stores, e.g., `prada.0`.)

```
$: git tag fhqwhgads
```

Having tagged the commit, we can now use the name `fhqwhgads` anywhere Git takes a commit ID.

If the commit we want isn't checked out right now, we can pass in a commit ID to tag:

```
$: git tag fhqwhgads 8891c37
```

Because nothing in Git can ever be simple, it turns out there are two kinds of tags. The kind we just created is a *lightweight* tag; it's stored in the repository as just a name pointing to a commit, similar to a branch.

The other kind is an *annotated* tag, which, in addition to a name and commit reference, can also include a message, similar to a commit message.

```
$: git tag fhqwhgads -a -m "Fhqwhgads release (22 »
   Dec 2014)"
```

Tags, like branches, can and should be shared on a remote, and you can push them to your remote the same way, using `git push`:

```
$: git push fhqwhgads
```

There aren't many rules surrounding tags, but the few rules that do exist are strict, as we'll see next.

Tag names must be unique

Just as it would be a huge problem if two different versions of your project could have the same name, Git does not allow you to create a tag if another tag by the same name already exists, and will reject a pushed tag if it already exists on the server.

Git will, however, let you give a tag the same name as a branch, or vice versa. But if you try to do anything ambiguous with a tag or branch name, Git will give precedence to the branch *and* will warn you that that may not have been the right move. Here's what happens when, in a repo that has both a tag and branch named branch-2, I try to check out branch-2:

```
$ git checkout branch-2
warning: refname 'branch-2' is ambiguous.
Switched to branch 'branch-2'
```

To make your life easier, avoid giving branches and tags the same names. A lot of teams who use tags will prepend something to their tag names to disambiguate them from branches; our `fhqwhgads` tag might instead be called `rel-fhqwhgads` to distinguish it from any `fhqwhgads` branches that may be flying around. This has the added benefit of saying what the tag refers to; in this case, `rel` is short for "release."

Tags are meant to be permanent

Git *will* let you change things like tags. More precisely, it will allow you to delete a tag and replace it with a new one under the same name. (To wit: if you do tag the wrong commit by accident, which sometimes happens, you can use `git tag -d <tagname>` to delete the bad tag and then create a new one pointing at the right commit.)

Having said that, a tag's purpose is to serve as a stable nickname for a specific commit—a job made more difficult if the names or commits underneath tags can change. Once you've pushed a tag to a remote—especially a remote you've shared with other people, like a collaborative hub—try never to change it. There may be times when you need to, or when re-creating a tag is simpler than creating a new tag with a new name, but I've found these situations to be exceptional, and not worth the headache of having to message your entire team to explain that `rel-wombat.0` may or may not *really* be the commit it's supposed to be.

TIME TRAVELING WITH git checkout

Reviewing what we've done is nice, but Git allows you to truly revisit the past by checking out old commits, using the same `git checkout` command you use to switch branches. I don't just mean "checking out" in the colloquial sense—"Hey, check out this cute panda video"—but in the version-control sense: when you check out a commit (or, for that matter, a branch), you're not just seeing a previous version of your work; you're *resetting* your local copy of the project to match whatever version you asked for. "Checking out" is used here in the same sense as a library book. And if it's unclear in this metaphor where your working tree fits in, remember that even if you're working progressively—adding new commits to a branch, rather than revisiting old ones—you still always have a version of the project checked out: the branch you're working on, to which you can add more commits.

Checking out a commit by itself differs from checking out a branch only in that you're not really expected to add any new commits after you check it out. That's not to say you *can't* add commits, though. To explain this distinction, let me give you an example.

Let's say you start getting reports from your users that something you *know* was working in a certain browser or device when you first deployed your project a few weeks ago is now no longer working. Let's also say that when you made that first production push, you also tagged the commit you pushed as `rel-v1.0`.

The first thing to do is confirm that the code you deployed originally actually did work, by checking out the old version and opening it up in a browser. Here we'll assume it's a static website that you can open directly in a browser, but if your site has a build step—using Grunt, Middleman, or some other tool—it should work here, too. Just run your build or server task after checking out the old site.

To do this, run `git checkout` with the tag or commit ID you want to return to:

```
$: git checkout rel-v1.0
Note: checking out 'rel-v1.0'.
```

This command did what we wanted it to do: Git has reset the files and folders in our working tree to match the version of our project we're trying to return to, which was commit `591672e`, also known by its tag, `rel-v1.0`. We can now open up the website and confirm that, yes, it worked when we shipped it. From here, we might continue our investigation by looking at the log, reviewing the commits that have been added to master since this one (`git log rel-v.10..master`), or look at the actual changes between this version and the latest one (`git diff rel-v.1.0..master`). If a particular commit seems likely to have introduced the bug, you can check it out to confirm (or allay) your suspicions. Git even offers a tool (`git bisect`) that performs this kind of binary search and automatically finds the commit that caused a particular issue. Tobias Günther wrote a great overview of using `git bisect` to squash bugs for *A List Apart* (http://bkaprt.com/gfh/05-11/).

What is the detached HEAD state?

When you check out a commit, as opposed to a branch, Git puts you into the "detached HEAD" state: your computer's HEAD pointer is pointed at a particular commit, but *not* at a branch. You're "detached" in the sense that you're not working on any branch. In practice, this means you can make new commits, and they will be saved, but you won't have a branch name to refer back to them.

You're not really supposed to add commits while detached. Most of the time, Git expects you to check out an old commit to review or test the old code, *not* to make changes. (That's what branches are for.)

But this can be a feature: commits made in the detached state can be used as a scratch pad. While detached, you're free to make experimental changes and commit them, and discard any commits you make in this state without impacting any branches by performing another checkout. There's even less risk than usual that a bad change will find its way into everyone else's copy of the project, or out to production, because (unless you move them into a branch with `git branch` or `git checkout -b`) commits in the detached state are homeless, unless you decide to create a branch to contain them. If you want to create a new branch to retain commits you create, you may do so (now or later) by using `-b` with the checkout command again. For example:

```
$: git checkout -b new_branch_name
HEAD is now at 591672e... Release v1.0
```

Returning from the detached HEAD state (reattaching the HEAD, so to speak) is as simple as checking out a branch—in this case, returning to master:

```
$: git checkout master
Previous HEAD position was 591672e... Release v1.0
Switched to branch 'master'
```

git checkout handles any reference that isn't a branch name as if it's just a commit, even if you're asking for it using a ref that includes the branch name, which can inadvertently lead you into the detached state. As we saw in the last chapter, if you use git checkout make-logo-bigger to check out the branch named make-logo-bigger, then you've checked out a branch. However, if you ask to check out origin/make-logo-bigger (a remote branch reference), you've checked out the commit that's currently at the head of that branch, but *not* the branch.

WRAPPING UP

One phrase that I've barely used in this book, outside of a few examples (where I've included it as an in-joke), is *directed acyclic graph* (DAG). A directed acyclic graph is a kind of data structure in which individual nodes point to other nodes, the references building on one another to form chains of information, spreading out like the roots of a tree, growing endlessly as we work, adding to the graph with every commit.

These kinds of graphs are often used to visualize Git branches, and it's not uncommon to see even the most basic Git tutorial include a bunch of branching diagrams.

To be fair, DAGs are a somewhat advanced concept, and most Git tutorials don't go so far down the rabbit hole as to mention them by name, even if they employ them as visual aids.

I mention DAGs here, as we wrap up our time together, to make a point about the philosophy of this book. When trying to explain Git, it's common to focus on the big picture: whole networks of repositories pushing and pulling one another, whole systems of branches flowing into and out of one another.

I'm not disparaging such attempts: these things are real, and they're spectacular. As a work of information science, I find Git incredibly beautiful.

But I also find that looking at these things as systems misses Git's most wonderful quality: people like you (and me, and our teammates) each making changes, evolving our projects one step at a time, crafting histories. The graph just isn't that important if what you're trying to do is save the next version of your project or share changes with your team. And although the history we collaborate on via Git can be modeled as a graph, it can also be a rendered as a list of incidents—as a story.

Admittedly, a Git repository is an odd place to tell a story; Git's command-line interface, not the most natural way to tell one. Back at the very beginning of this book, I described Git's interface as a "leaky abstraction." Git tries, but doesn't always succeed, to protect us from having to understand the many complex things going on when we run a particular command. In not succeeding, Git encourages us to learn about what's actually going on behind the scenes.

But the stories we tell together are just as real and beautiful as the information structures the creators of Git have created to contain them. And now, I hope, you'll be armed with the knowledge to tell these stories with a minimum of fear.

CONCLUSION

IN THIS BOOK, we've covered everything from the difficulty of revising writing carved in stone to tips for how best to take advantage of a detached HEAD. Along the way, we've learned a few things about Git: what commits are made of, how each commit is a whole version of your work, and how commits, along with remotes, branches, and other stuff, come together to create a wild new landscape of things that—good news!—you now need to worry about in your daily work. I'll understand if you've ended up here, at the end of not-the-shortest book in the A Book Apart series, still harboring at least a few questions about Git.

That's okay! We've only scratched—by design—the surface of what Git is capable of. It's less important for you to come away from this book knowing every single Git command than it is for you to know how Git thinks and, from there, to understand that Git is neither evil, nor magical, nor scary. It's just a tool and, if you use it properly, it will always serve you well.

More than that, though, you can use the commands and functions we've covered in this book as building blocks for finding your own satisfying Git workflows, and as jumping-off points for learning new tricks. Depending on the kind of work you do, you'll either find that the knowledge imparted by this book is more than enough to help you get the job done, or you'll feel equipped to ask more incisive questions about how Git can better serve you in the future.

RESOURCES

Command Reference

Here's a quick list of every Git command referenced in this book, plus a few others. Arguments in square brackets (e.g., [thing]) are optional.

git config [--global] <key> <value>

Updates Git's settings, modifying the preference identified by <key>, such as user.email, with the given <value>, such as david@demaree.me. The --global flag saves preferences to a file in your home directory, so Git will apply them to every project on your computer. Otherwise they're saved and applied only within a specific project.

git init

Creates a new Git project inside the current working directory—that is, if you're inside a directory named my-awesome-project that contains a website you're working on, running git init will turn the folder into a fresh Git repository, ready to use.

git clone <url> [directory]

Copies an existing Git project located at the given url to your computer as a new directory. By default, the directory will be named after the Git repository in the URL—the repo https://gitforhumans.info/rails.git would be copied into a folder named rails, but you can provide your own directory name as an argument if you want.

git status [-s] [path/to/thing]

Outputs the status of your working copy: identifies which files are modified but not staged, or added but not committed. The optional --short or -s flag gives you a shorthand version of the status readout. By default, git status will show you the status of everything in your project, but you can give it a directory or file path to limit the results.

git add [--all] filename.txt

Adds a changed file to the staging area for inclusion in the next commit.

git rm folder/filename.txt

A shortcut command that deletes the file at the given path, then stages the deletion for your next commit. If you've already deleted the file elsewhere (say, via the Finder), it just stages the change.

git mv oldpath.txt newpath.txt

Another shortcut that moves the file at `oldpath.txt` to `newpath.txt`, then stages that change.

git reset filename.txt

The opposite of `git add`: having staged a change to `filename.txt`, you can use `git reset` to *un-*stage it.

git commit [-a] [-m "Your message"]

Adds a commit with any changes you've staged using `git add`. The `--all` (or `-a`) option is a handy shortcut—it will automatically stage any changes you've made to your working copy. You can use the `--message` (`-m`) argument to specify your commit message; if left blank, Git will open up your default text editor (or whatever editor you've configured in Git's settings).

git branch [-r|-a]

Shows a list of all your branches. By default, it shows you only branches on your local copy of the repo. The `-r` option can show you all the branches you've fetched from remotes; `-a` shows you both local and remote branches.

git branch ‹branchname› [‹commit›]

If you give a branch name as an argument to `git branch`, it'll create a branch with that name, starting at the current commit (or at any commit you specify, if you provide its ID).

git checkout [-b] ‹branchname-or-commit›

Updates your working copy to match the given branch or commit—in essence, switching you into that branch/commit. If you check out a branch, Git sets that as the current branch so you can add commits to it. If you check out a commit or tag, Git "detaches" from any branch—you can make commits, but they will only be retrievable by their commit IDs.

git merge ‹otherbranch›

Merges `otherbranch` into the current branch, provided there are no conflicts. If there are conflicts, Git copies over and stages as much of what's in the other branch as possible, marking the conflicted files so you can resolve the problem yourself before committing.

git remote add ‹name› ‹url›

Adds a remote with the given name and URL to your local Git project settings.

git remote rm ‹name›

Removes the remote from your project settings along with any remote tracking branches you may have fetched from the server. Note that this only deletes the remote from your local settings—everyone else's computers, and the server, are not affected.

git push <remotename> <branchname>

Pushes the current state of branchname to the remote named remotename.

git pull <remotename> <branchname>

Pulls down the current state of branchname from the remote to your local copy, and attempts to merge it into your current branch.

git fetch <remotename>

Copies everything from the remote to your local copy. When you run git pull, a fetch happens automatically.

git log [--oneline] [--pretty] [<branchname-or-commit>]

Shows a reverse-ordered list of commits, starting from the current head (or any one you specify by branch name or ID). You can use the --pretty option to customize the output; --oneline is a shortcut for the most used output format, consisting of a short commit ID and the commit message on each line.

git diff [--stat] [<branchname-or-commit>]

Generates a "diff"—a visual representation of the differences between two commits. The --stat option produces a summary view showing a list of files changed, with how many lines were added and deleted in each one.

git tag [-a] [-m] <tagname> [<commit>]

Tags a commit with the name you provide, which you can use as a static, friendly name for that commit. The -a flag tells Git to create an annotated tag, which includes information about when the tag was created, by whom, and a message saying what it's about, just like a commit. (Otherwise, Git creates a "lightweight" tag, which references a commit but doesn't create any of that other info.) If you create an annotated tag, make sure to include the `--message`/`-m` argument, again, just like a commit.

git tag -d ‹tagname›

You shouldn't need to delete a tag, but if you do, you can do it by passing the `-d` (for "delete") option to `git tag`.

git tag -l

Outputs a list of all the tags in your repository.

git push --tags ‹remotename›

As a safeguard against accidentally sharing a tag that you might not be ready to share, Git doesn't push any of your tags unless you include the `--tags` option.

For an exhaustive list of all Git's commands and complete details on how to use them, check out the documentation on Git's website (http://bkaprt.com/gfh/06-01/).

Recommended Git apps

In this book I've chosen to focus on Git's command line interface in order to best demonstrate how Git thinks, and I still recommend that you start with the command line. However, there are many excellent time-saving Windows and Mac apps you can use once you're up and running.

GitHub Desktop. Whether or not you host your code on GitHub, their desktop apps for Mac and Windows are among the very best— and they're free. You can visually stage and commit changes, create and switch between branches, push and pull with remotes, and if you *do* host on GitHub, the desktop app makes it easy to create pull requests or open a compare view (http://bkaprt.com/gfh/06-02/).

Tower. For Mac power users willing to spend $70, Tower offers many more options and features. Where GitHub Desktop focuses on the basics, Tower can also handle resolving merge conflicts, cherry-picking commits, and lots more (http://bkaprt. com/gfh/06-03/).

SourceTree. More complex and powerful than GitHub's apps, but lacking some of Tower's slickest features, SourceTree (which is free) is a good choice for someone who wants a little more power in a Git app, but doesn't want to spend money (http://bkaprt.com/gfh/06-04/).

Many popular coding tools also include built-in support for Git, or allow you to add it via plugins, so you can commit changes without leaving the app. Atom, Coda, Sublime Text, TextMate, BBEdit, Xcode, and Visual Studio Code all work with Git out of the box.

Git hosting services

GitHub. The biggest—and, at one time, kind of the *only*—name in Git hosting. Chances are, if you work with code you've had to do *something* on GitHub, because it's what everyone uses. Ubiquity aside, GitHub remains arguably the best choice for most people: the company continues to invest in tools and resources that make it easier to collaborate via Git (such as Pull Requests), as well as new and interesting tools like GitHub Pages (web hosting powered by a Git repo). GitHub charges money to host private projects for yourself or your organization. Public projects where anyone can pull or download your code, but only you and your teammates can push changes, are always free. There's also an enterprise edition that costs *lots* of money, but you can run it on your own servers for maximum control over your data (github.com).

Bitbucket. Not as slick as GitHub, Bitbucket has one nice benefit for hobbyists or small businesses: individuals and small teams can host unlimited private repos for free. Although Bitbucket lacks GitHub's vast community, I personally use both: GitHub for public projects or collaborative work, Bitbucket for small personal projects (bitbucket.org).

Beanstalk. Specializing in paid, private repos, Beanstalk has a few nice features for web developers, most notably a built-in deployment tool that automatically updates your web servers after new code is pushed to your repository (beanstalkapp.com).

Finally, if you're handy with the command line and either need to have total control over your data or enjoy a bit of extra nerdery, it's not that hard to roll your own hosting. Because Git's default protocol is SSH, any Linux server can conceivably be set up to host Git repositories. The folks at DigitalOcean have a handy guide to setting up a simple Git server on one of their virtual servers (http://bkaprt.com/gfh/06-05/).

Even more Git

"Meet the Command Line", Hosted by Dan Benjamin, these screencasts are a great resource for command-line novices, allowing them to quickly get comfortable talking to a computer via the Terminal (http://bkaprt.com/gfh/06-06/). Speaking of screencasts, Pluralsight's Git tutorials by James Kovacs cover everything from installing Git to more advanced branching and rebasing topics (http://bkaprt.com/gfh/06-07/).

GitHub Training. This microsite offers two online "courses" in slide-show format, one for command-line veterans and another for graphical app users. It also offers a PDF quick reference covering many of the basic commands, and even a sweet browser-based Git simulator to help practice typing commands (http://bkaprt.com/gfh/06-08/).

Atlassian's Git tutorials. An excellent resource for new and veteran Git users alike, these tutorials cover a range of topics with clear writing and illustrations. In particular, their guide comparing different Git workflows is a great, deeper dive into the hub model I talk about in this book (http://bkaprt.com/gfh/06-09/).

Git for Type Designers. Frank Grießhammer has put together a great quick reference to basic Git workflows, with specific notes for using Git with type design tools (http://bkaprt.com/gfh/06-10/).

Pro Git. Scott Chacon's open-source book is the definitive reference for humans who want to truly understand how Git works (http://bkaprt.com/gfh/06-11/).

Notable Git repos

A Git repo tracking the complete German legal code. As its maintainers write, "all German citizens can easily find an up-to-date version of their laws online. However, the legislation process, the historic evolution, and the updates to laws are not easily and freely trackable." By publishing a copy of the laws via Git, anyone can see how laws change over time by simply reviewing the commit log (http://bkaprt.com/gfh/06-12/).

Vox Media's Code of Conduct. Vox has open-sourced their code of conduct, a living document that, Mandy Brown writes, "will evolve and grow with our team as well as with input from the community." The source code for the code-of-conduct website is hosted on GitHub; members of the community can file issues, submit patches, or simply review the commit log to see the code's evolution over time (http://bkaprt.com/gfh/06-13/).

Tacofancy. A "community-driven, object-oriented taco recipe repo" created and maintained by Dan Sinker, Tacofancy offers complete recipes for tacos, as well as individual taco components (http://bkaprt.com/gfh/06-14/).

What is Code? In 2015, writer, programmer, and prolific GitHub user Paul Ford (http://bkaprt.com/gfh/06-15/) wrote a Bloomberg Businessweek article titled "What is Code?" (http://bkaprt.com/gfh/06-16/), an epic explainer covering, among other things, GitHub and Git. The source code for the article's playfully interactive web version is publicly viewable on GitHub (http://bkaprt.com/gfh/06-17/).

ACKNOWLEDGMENTS

This book wouldn't exist without help and encouragement from the incredible editorial and production team at A Book Apart. Even as her author became a stressed, sleep-deprived new parent, Katel LeDû kept things moving with remarkable poise. Caren Litherland's editorial guidance helped *Git for Humans* find its shape, and helped turn a pile of nonsense into something resembling a book. Jason Santa Maria, Rob Weychert, and Ron Bilodeau made my words look beautiful and readable on pages and screens. And, last but not least, thanks to Jeffrey Zeldman for getting all of us into this mess.

High fives and whiskey to David Yee, our tech editor, who asked questions and flagged confusing bits with smarts, empathy, and just the right amount of pedantry.

My deepest thanks (and more whiskey) to Mandy Brown, for seeing a book in me, offering me the chance to write this one for A Book Apart, and, more than two years later when it was finally finished, writing the foreword.

Thanks to friends and colleagues who offered great feedback on early drafts and much-needed encouragement, like asking "is the book done yet?": Tim Brown, Paul Hammond, Morgan Kelly, Sally Kerrigan, Liz Galle, Jake Giltsoff, Lucy Knisley, Bram Stein, Kevin Stewart, and Elliot Jay Stocks.

To every designer whose complaints about Git in my Twitter feed provided the inspiration for this book, particularly Ethan Marcotte, Frank Chimero, Susan Robertson, and especially Mat Marquis and Dave Rupert, who were the first to ask in so many words for A Book Apart to publish a book like the one you're holding—I wrote this for all of you.

To the gits behind Git & GitHub: Linus Torvalds, Junio Hamano, Scott Chacon, Chris Wanstrath, PJ Hyett, Tom Preston-Werner, and countless others who've contributed code, words, or both, so that we can share our code and words with each other: all of this is your fault. Thank you.

To Jeff Veen, Bryan Mason, Ryan Carver, Greg Veen, Matthew Rechs, and the whole Typekit team, for showing me how to tackle the hard problems, and then putting up with me for the last two years while I tackled this one.

Finally, my deepest love and gratitude to my wife Jody for her love, patience, and support, and to our baby daughter June, who is so looking forward to eating this book.

REFERENCES

Shortened URLs are numbered sequentially; the related long URLs are listed below for reference.

Introduction

00-01   http://frankchimero.com/writing//two-sentences-about-getting-older-and-working-on-the-web/

00-02   http://www.joelonsoftware.com/articles/LeakyAbstractions.html

Chapter 1

01-01   http://www.pcworld.idg.com.au/article/129776/after_controversy_torvalds_begins_work_git_/

01-02   https://github.com/torvalds/linux

Chapter 3

03-01   https://en.wikipedia.org/wiki/Exquisite_corpse

03-02   http://www.layertennis.com/

Chapter 4

04-01   https://help.github.com/articles/generating-ssh-keys/

04-02   https://git-scm.com/book/ch4-1.html

Chapter 5

05-01   http://git-scm.com/docs/pretty-formats

05-02   https://www.atlassian.com/git/tutorials/git-log/formatting-log-output

05-03   https://www.atlassian.com/git/tutorials/inspecting-a-repository/git-log/

05-04   https://www.youtube.com/watch?v=4XpnKHJAok8&t=56m20s

05-05   https://help.github.com/articles/associating-text-editors-with-git/

05-06   http://tbaggery.com/2008/04/19/a-note-about-git-commit-messages.html

05-07   http://chris.beams.io/posts/git-commit/

05-08   http://24ways.org/2014/dealing-with-emergencies-in-git/

05-09 http://www.kaleidoscopeapp.com/

05-10 https://github.com/rails/rails/tree/v4.2.0

05-11 http://alistapart.com/article/git-the-safety-net-for-your-projects

Resources

06-01 https://git-scm.com/docs

06-02 https://desktop.github.com

06-03 http://www.git-tower.com

06-04 http://sourcetreeapp.com/

06-05 https://www.digitalocean.com/community/tutorials/how-to-set-up-a-private-git-server-on-a-vps

06-06 https://www.pluralsight.com/courses/meet-command-line

06-07 https://www.pluralsight.com/courses/git-fundamentals

06-08 http://training.github.com/

06-09 https://www.atlassian.com/git/tutorials

06-10 http://git-scm.com/book

06-11 https://github.com/bundestag/gesetze

06-12 https://github.com/voxmedia/code-of-conduct

06-13 https://github.com/sinker/tacofancy

06-14 https://github.com/ftrain

06-15 http://www.bloomberg.com/whatiscode

06-16 https://github.com/BloombergMedia/whatiscode

INDEX

ABOUT A BOOK APART

We cover the emerging and essential topics in web design and development with style, clarity, and above all, brevity—because working designer-developers can't afford to waste time.

COLOPHON

The text is set in FF Yoga and its companion, FF Yoga Sans, both by Xavier Dupré. Headlines and cover are set in Titling Gothic by David Berlow.

ABOUT THE AUTHOR

David Demaree is a web developer, designer, speaker, and product person based just outside New York City. He's a senior product manager for Adobe Typekit, working on ways to make it easy for everyone to find and use great fonts wherever they need type. David has spoken at design and tech events in the United States, Europe, and Australia, and he writes about software on Medium.

Photograph by Ryan Carver

www.ingramcontent.com/pod-product-compliance
Lightning Source LLC
Chambersburg PA
CBHW040900210326
41597CB00029B/4912